Stereo Vision for Facet Type Cameras

DISSERTATION

zur Erlangung des Grades eines Doktors
der Ingenieurwissenschaften

vorgelegt von
M.Sc. Tao Jiang

eingereicht bei der Naturwissenschaftlich-Technischen Fakultät
der Universität Siegen

Siegen 2013

stand: October 2013

Bibliographic information published by the Deutsche Nationalbibliothek

The Deutsche Nationalbibliothek lists this publication in the Deutsche Nationalbibliografie; detailed bibliographic data are available on the Internet at http://dnb.d-nb.de .

ISBN 978-3-8325-4285-6

Logos Verlag Berlin GmbH
Comeniushof, Gubener Str. 47,
10243 Berlin
Tel.: +49 (0)30 42 85 10 90
Fax: +49 (0)30 42 85 10 92
INTERNET: http://www.logos-verlag.de

Gutachter der Dissertation:
Prof. Dr. Klaus-Dieter Kuhnert
Prof. Dr. Marcin Grzegorzek

Datum der mündlichen Prüfung:
27, September 2013

Acknowledgements

During my times as a Ph.D. student at the University of Siegen I have had the fortune of knowing and working with many amazing people. Your help, support and friendship has made my time here truly incredible.

First of all, I would like to thank my supervisor Professor Dr. Klaus-Dieter Kuhnert for his constant guidance and support over the years. His wisdom, humor, and kindness are truly inspiring. He brought me into the field of computer vision and taught me the principles of scientific work and writing. When my research had suffered some setbacks, his encouragement let me persevere in my efforts. Without his careful guidance, this work would not have been possible. A special thanks to Professor Dr. Marcin Grzegorzek, who was willing to take on the duty of this thesis' second referee.

Also, I would like to thank my thesis committee. They spend a lot of their valuable time to read and remark my thesis, which guided and improved my thesis. I want to express my gratitude to my colleagues at the Institute of Real-time Learning Systems for their cooperation and help, especially Duong Nguyen-Van, who always discussed with me and gave me a great source of inspiration. Thanks to Lars Kuhnert and Markus Ax who taught me about the robots'operation and helped me to resolve many technical problems. I am grateful to Stefan Thamke and Jens Schlemper who kindly provided me all kinds of support and help in my work and living. I would also like to thank Klaus Müller, Ievgen Smielik and Sailan Khaled for their encouragement, enthusiasm and support.

During my studies, I had the pleasure to supervise the projects and the master theses of Maoxia Hu and Tao Ma. Their work proved to be of enormous value for my research. I want to thank them for their contribution in my experiments. I would like specially to thank Dr. Frank Wippermann and Dr. Alexander Oberdoerster who work in the German Fraunhofer Institute for Applied Optics and Precision Engineering. They had not only provided a demonstration system of eCley for my experimental research but also given me much useful instruction and valuable suggestion. Without their help, the real development of eCleys' stereo vision would have been much more difficult and time-consuming. I would also like to acknowledge the China Scholarship Council (CSC) for financing my research and living.

I want to express my gratitude to all of my friends and people who supported me, especially Professor Gexiang Zhang. I have measured myself against each of him as a model for outstanding academic and professional success. I would also like to thank my family, especially my mother and father, who gave me so much, and helped me become who I am today. I thank my wife Lingmei Chen for being at my side throughout all these years. She never has any complaint about the many late nights at the lab and constantly provided comfort during difficult times. I am grateful to my daughter Peiran who always gave me happiness and liveliness.

This thesis has also been funded by the graduation aid program of the DAAD and Institute of Real-time Learning Systems at University of Siegen, by the Scientific Research Foundation of CUIT (No. KYTZ201410, J201508), by Scientific Research Fund of SiChuan Provincial Science & Technology Department (No. 2015GZ0304), and by Scientific Reserch Fund of SiChuan Provincial Education Department (No. 16ZA0206). Their support is gratefully acknowledged.

Abstract

In the last decade, scientists have put forth many artificial compound eye systems, inspired by the compound eyes of all kinds of insects. These systems, employing multi-aperture optical systems instead of single-aperture optical systems, provide many specific characteristics, such as small volume, light weight, large view, and high sensitivity. Electronic cluster eye (eCley) is a state-of-the-art artificial super-position compound eye with super resolution, which is inspired by a wasp parasite called the Xenos Peckii. Thanks to the inherent characteristics of eCley, it has successfully been applied to aspects of medical inspection, personal identification, bank safety, robot navigation, and missile guidance. However, all these applications only involve a two-dimensional image space, i.e., no three-dimensional (3D) information is provided. Conceiving of the ability of detecting 3D space information using eCley, the performances of 3D reconstruction, object position, and distance measurement will be obtained easily from the single eCley rather than requiring extra depth information devices.

In practice, there is a big challenge to implementing 3D space information detection in the minimized eCley, although structures similar to stereo vision exist in each pair of adjacent channels. In the case of an imaging channel with short focal length and low resolution, the determination of the depth information not only is an ill-posed problem but also varies in the range of one pixel from quite near distance ($\geq 86mm$), which restricts the applicability of popular stereo matching algorithms to eCley.

Taking aim at this limitation, and with the goal of satisfying the real demands of applications in eCley, this thesis mainly studies a novel method of subpixel stereo vision for eCley. This method utilizes the significant property of object edges still retained in eCley, i.e., the transitional areas of edges contain rich information including the depths or distances of objects, to determine subpixel distances of the corresponding pixel pairs in the adjacent channels, to further obtain the objects' depth information by employing the triangle relationship. In the whole thesis, I mainly deduce the mathematical model of stereo vision in eCley theoretically based on its special structure, discuss the optical correction and geometric calibration that are essential to high precision measurement, study the implementation of methods of the subpixel baselines for each pixel pair based on intensity information and gradient information in transitional areas, and eventually implement real-time subpixel distance measurement for objects through these edge features.

To verify the various methods adopted, and to analyze the precision of these methods, I employ an artificial synthetical stereo channel image and a large number of real images captured in diverse scenes in my experiments. The results from either a process or the whole method prove that the proposed methods efficiently implement stereo vision in eCley and the measurement of the subpixel distance of stereo pixel pairs. Through a sensitivity analysis with respect to illumination, object distances, and pixel positions, I verify that the proposed method also performs robustly in many scenes. This stereo vision method extends the ability of perceiving 3D information in eCley, and makes it applicable to more comprehensive fields such as 3D object position, distance measurement, and 3D reconstruction.

Zusammenfassung

Ausgehend von den Facettenaugen der Insekten haben Wissenschaftler seit 10 Jahren viele künstliche Facettenaugensysteme erstellt, die auf der Multi-Apertur-Optik basieren. Im Vergleich zu den auf Single-Apertur-Optik basierenden Systemen sind diese Systeme kleiner und leichter. Außerdem haben solche Systeme ein großes Sichtfeld und eine hohe Empfindlichkeit. Das eCley (Electronic cluster eye) ist ein neues künstliches Facettenaugensystem, das Bilder mit Super-Pixel-Auflösung erstellen kann, welches vom Sehsystem der parasitären Wespe „Xenos Peckii" inspiriert ist. Wegen seiner ausgezeichneten Fähigkeiten sind eCley-Systeme in den Bereichen ärztliche Untersuchung, Identitätsauthentifizierung, Roboternavigation und Flugkörperlenkung angewendet worden. Aber solche Anwendungen basieren nur auf der Datenverarbeitung im 2D-Bereich. Wenn jedoch mit einem eCley-System räumliche 3D-Daten erzeugt werden können, kann man nur mit eCley 3D-Rekonstruktion, Lokalisierung und Entfernungsmessung erledigen, die man vorher mit anderen Geräten durchführen musste.

Zwar können je zwei horizontal benachbarte Mikrokameras im eCley als ein Stereo-Sehsystem genutzt werden, aber es ist nicht leicht, die räumlichen Informationen durch so kleine Kameras zu erhalten. Die von der Mikrokamera gemachten Fotos haben nur eine ziemlich niedrige Auflösung. Außerdem ist die Tiefenveränderung der Szene kleiner als 1 Pixel, wenn die Entfernung größer als 86mm ist, d.h. dass viele verbreitete Algorithmen zum Stereosehen mit eCley nicht gut funktionieren können.

Um die verbreiteten Stereosehalgorithmen mit dem eCley besser anwenden zu können, wurde eine neue Methode dafür im Bereich des Subpixel-Stereosehen erstellt. Diese Methode basiert auf der positiven Eigenschaft des eCleys, dass die Kanten des Ziels im eCley sehr gut behalten werden können. Im bergang zwischen Bilder benachbarter Mikrokameras gibt es zahlreiche Tiefeninformationen. Mit diesen Tiefeninformationen kann der entsprechende Subpixelabstand ausgerechnet werden. Danach kann die Entfernung des Ziels mit dem Subpixelabstand berechnet werden. Aufgrund der Struktur des eCleys haben wir in dieser Doktorarbeit ein mathematisches Modell des Stereosehens für eCley abgeleitet. Dazu werden die optische Ausrichtung und die geometrische Korrektur, die die Voraussetzungen zur präzisen Messung sind, diskutiert. Zum Schluss haben wir die Subpixel-Baseline-Methode, die auf der Helligkeit und den Gradienten basiert, und die Echtzeit-Messung für den Subpixelabstand, die auf der Eigenschaft der Kanten basiert, entwickelt.

Um unsere Methode zu überprüfen, haben wir viele künstliche und reale Szenenbilder angewendet. Das Ergebnis zeigt, dass unsere Methode die Messung zum Subpixelabstand für Stereopixelpaare ausgezeichnet realisiert hat. Außerdem funktioniert diese Methode in vielen komplexen Umgebungen robust. Das bedeutet, dass die Methode die Fähigkeit des eCleys verbessert hat, die 3D-Umgebung zu erkennen. Das eCley kann daher in verschiedenen 3D-Anwendungsbereichen eingesetzt werden.

Contents

CONTENTS

List of Figures

Chapter 1

Introduction

The acquirement of depth information is quite important for the artificial compound eye: it is a prerequisite for object measurement, object tracking and 3D reconstruction. In this chapter, the background for the artificial compound eye and the present study are introduced first. Then the motivation is given after analyzing the demand for applications, which is followed by an outline of the whole structure of the thesis.

1.1 Artificial Compound Eye

1.1.1 General Artificial Compound Eye

Nowadays, optic imaging systems are broadly applied to various tasks, and the demand for compact vision systems is growing rapidly. A miniaturized imaging system has the advantages of low energy consumption, small volume, and a large field of view (FOV) so that more and more researchers are focussing on it to broadly extend its applicability. At present, these miniaturized imaging systems have been successfully applied to personal mobile phones. Further on, they may be expected to be implemented in smaller objects such as credit cards. Thus, the size of the system volume is required to be smaller and smaller. When miniaturizing a classical vision system based on the single aperture eye, one encounters the limitation of diffraction [Duparr & Wippermann, 2006]. So, researchers have started paying more attention to the development of the artificial compound eye. The artificial compound eye combines hundreds and thousands of ommatidia, also called optical channels, and simulates the compound eye of creatures such as the fly, honeybee, dragonfly, wasp, etc., to reach the desired goals of small volume, light weight, and a large field of view [Duparr & Wippermann, 2006][Nakamura *et al.*, 2012]. At present, the artificial compound eye has successfully been applied to lithographic imaging [Voelkel *et al.*, 1996], intelligent robot vision [Tong *et al.*, 2011], missile guidance [Duparr *et al.*, 2005b], and some applications in civil industry such as large-scale infrared telescopes, mini cameras, and fingerprint identification systems [Duparr & Wippermann, 2006].

In general, based on the natural compound eyes of organic creatures, the artificial compound eyes for compact vision systems can be divided into two classes: apposition and superposition compound eyes [Duparr & Wippermann, 2006]. The apposition compound eye consists of an array of microlenses and a pinhole array on the front and backside of the spacing structure respectively. Each microlens is associated with a small group of photoreceptors that are formed by the pinhole array in its focal plane, and the single microlens receptor unit forms one optical channel, also referred to as an ommatidium [Duparr et al., 2005a]. The superposition compound eye, also called the cluster eye, is more complex than the apposition one, and has the arrangement of its refracting surfaces similar to that of a Gabor superlens [Stollberg et al., 2009] in order to achieve an image merger, and has field apertures in the intermediate image plane to avoid overlay of different channel images and to reduce aberration. The light from multiple facets combines to form a single erect image of the object on the surface of the photoreceptor layer [Nakamura et al., 2012].

Compared to the apposition compound eye, the superposition compound eye is much more sensitive to light, and so achieves a higher resolution. In practice, when miniature focal length and volume are required, the artificial apposition compound eye can be considered as a simple imaging optical sensor whereas the cluster eye is capable of becoming a valid alternative to classical bulk objectives. The vision of the artificial compound eye can currently achieve a large field view of $92°$ and a thickness of 0.3 mm.

1.1.2 Electronic Cluster Eye (eCley)

Recently, a new artificial superposition compound eye, called the electronic cluster eye (eCley) is being developed by the German Fraunhofer Institute for Applied Optics and Precision Engineering, to achieve super-resolution. It is inspired by a wasp parasite called Xenos Peckii [Brueckner et al., 2010]. As to the eCley, each channel only records a part of the whole FOV, i.e. a cluster of small partial images, and creates a final image through the stitching together of all the partial images by means of software processing. So a compact size, a large FOV and a good sensitivity in the visible spectral range are achieved together.

The eCley is shown in Figure 1.1. Figure 1.1(a) demonstrates the micro lens array with 17 by 13 units. The whole micro lens array is closely attached to the sensor, which definitely decreases the focal length to be smaller than 1 mm. Figure 1.2 shows the channel images that capture the parts of the same scene through each micro lens and the stitched image with high resolution.

Since the eCley has the property of compact size, a large FOV, and convenient assembly,

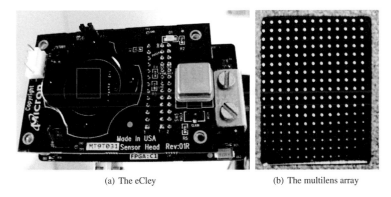

(a) The eCley   (b) The multilens array

Figure 1.1: Artificial compound eye: the eCley

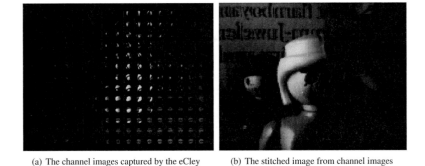

(a) The channel images captured by the eCley   (b) The stitched image from channel images

Figure 1.2: The channel images and real scene image captured by the eCley

the eCley can be widely applied to the inspection of surfaces, bio-medical imaging, document analysis, digitalization of photographic film material, and so on. However, most applications do not fully exploit the properties of the eCley such as its short focal length, and there are still more potential functions, such as measuring distances, which need to be developed so that the eCley can possess powerful and integrated performances in real applications. The eCley has an inherent property analogous to that of the stereo camera, that is, each channel and its adjacent channel comprise a similar pair of a small stereo camera. However, the short focal length and tiny channel image limit the function of stereo matching, or more accurately, the perception of object distances. How can one break through this barrier and enable the eCley to perceive the distances of objects?

3

1.2 Motivation and Task

1.2.1 Motivation and Goal

Artificial compound eyes possess many special properties, such as parallel multiple lenses, which facilitate their application to many military and industrial fields where common cameras are not normally competent. Early on, the polarization navigator was applied to sailing. It arranges the polarization units along with each ommatidium according to the structure of the visual cells of bee eyes [Kagawa et al., 2012]. Using this compound eye to see the sky, the direction of polarized light is derived from the deviation of patterns regarding the position of sun. Based on the property of multiple apertures, the hemisphere optical device with multiple apertures has been designed to search for objects, such as infrared telescopes of large scale, alarm satellites, and space monitoring systems [Xianwei et al., 2013]. Salient functions of an insect's compound eyes include the capacity to precisely process the vision information, calculate in real-time the position and direction of tracked objects, simultaneously issue commands to control the fly's direction and speed, and track and intercept objects. In a similar way, precision-guided imaging in the military field has been a successful application of the artificial compound eye, utilizing two reticles and an imaging sensor to construct a multimode guided device in a missile head to obtain the 3D information of objects [Li & Yi, 2010]. This guide head not only inherits the properties of a large VOF, light weight, small volume, and short focal length, but also is strongly fault-tolerant, adaptive, and highly tracking-capable. Duparré et al. developed a full FOV imaging system based on an artificial apposition compound eye, which is quite suitable for detecting small holes and the use in medical endoscopes. In this system, the optical information can be received and form image pieces at different views by microlenses that connect with a special mechanical rotation axis, and then each image piece can be combined into a full FOV image [Duparr et al., 2007].

Depth information is significant in the field of computer vision, which is the essential ingredient for many real applications like robot navigation, distance measurement, object position, three dimensional reconstruction, and so on. Similarly, if the eCley were to become capable of getting depth information, that would greatly improve its performance and extend its field of application to areas such as small object searching, micro distance measurement, fingerprint reconstruction, and many more. To date, some research has been devoted to developing the ability to obtain 3D information based on some kind of artificial compound eye. Horisaki et al. [Horisaki et al., 2007] proposed an approach to three-dimensional information acquisition using a compound imaging system based on the modified pixel rearrangement method and the estimation of the object distance. This method utilizes multiple images observed from different viewpoints, which are captured by a kind of artificial compound eye, the so called TOMBO (thin observation module by bound optics), to determine the depth information and reconstruct high-resolution images from multiple low-resolution images rather than only improve the resolution of the reconstructed image. However, that is only suitable for the TOMBO rather than the eCley, where constant offsets exist between two adjacent channels. Unfortunately, the functionality of depth information has not yet been implemented in the eCley. With the increasing demand for miniaturized cameras and for extensions to the field of applications, the functionality of depth information, i.e. object distance measurement, is becoming more and more significant to the eCley, which inevitably affects its developmental prospects. I expect that the eCley could attain a similar ability as that of a stereo camera to obtain disparities. This would greatly improve the eCley's performance and value. This thesis is dedicated to implementing 3D distance measurement using the eCley.

1.2.2 Main Contributions

The goal of this dissertation is to implement distance measurement for the eCley in a real environment. Our research demonstrates the inherent structure of an artificial compound eye, the intensity properties of the transitional areas, its optical and geometric calibration, the detection of corresponding pixel points, subpixel baseline, and subpixel distance measurement. The main contributions addressing key technical issues in subpixel distance measurement are listed as follows:

- Based on the special structure of the eCley, I analyze the inherent prerequisites for stereo vision in the eCley, which contain the same optical parameters and parallel imaging plane for each channel. I also discuss the limitations of implementing the usual stereo vision.

- Exploiting the property that the transitional areas of edges embody gradual changes of intensity from the object to the background, i.e. the edges contain distance information about the object, I demonstrate relation of the subpixel positions of the edges with intensity changes in the transitional areas, and then propose that the disparity between a pair of stereo corresponding points in two adjacent channels can be determined from the differences in the subpixel positions of edge points having the same intensity.

- I propose two brightness correction methods: one based on the maximum of the intensities and the other based on the voted probability. With respect to deriving the offsets of the channels, I use circle fitting to the subpixel coordinates of circle centers, further improving the accuracy of the subpixel coordinates through line fitting. Additionally, the radial and tangential distortions are removed by the least squares method and a rectification of the two channels is also implemented.

5

- With regard to the ill-posed problem of edge processing, I propose two approaches to obtain the positions of corresponding points in adjacent channels. One is to detect transitional areas according to their intensity changes and then select the positions of points with respect to the mean values of two transitional areas in the channel images. The other is to detect the gradient range of edges roughly at the pixel level, and derive the subpixel positions of the edges from fitting a sigmoid function to the edges. The proposed methods can efficiently and accurately determine the subpixel distance between two corresponding points of an object's edges, i.e. calibrate the subpixel baseline in practice.

- In a real environment, I give a real-time method of measuring the subpixel distances of objects, which first employs the edge detector to obtain the position of the edges and transitional areas, and then interpolation methods are applied to determine the fine subpixel disparities. This method not only attains highly accurate results, i.e. accurate distances of objects, but also ensures that the processing can be accomplished in real time.

1.2.3 Organization of the Dissertation

This dissertation contains seven chapters, which describe in detail the main research involved with the contributions. The remainder of this document is organized as follows:

- Chapter 2 presents the situation of stereo vision in the usual cameras, the diversity of stereo matching algorithms categorized into local and global methods, and the approach to 3D measurement via canonical stereo vision. It then discusses the depth-resolution sensitivity. In addition, the current 3D information function in the artificial compound eye and the limitations of general stereo vision for the eCley are stated in detail.

- Chapter 3 discusses the need to process images at the subpixel level and the implementation methods of subpixel resolution. It is followed by an introduction to the stereo vision structure of a multilens camera. I demonstrate a significant property: the intensity change in the transitional area between objects and the background contains position information which is still retained in the channel images. Then an overview of the implementation methods is summarized by a diagram.

- Chapter 4 describes how to perform brightness corrections based on the methods of maximum value and voted probability in the transitional areas. Then I propose a method of determining the subpixel coordinates of channel centers by fitting circles, and how to further improve the accuracy of these outcomes by the constraint of co-linear centers in each row or column, i.e. straight line fitting. The accurate subpixel parameters of the centers can be used to derive the offset between the two adjacent channels. I also establish

6

the distortion model including radial and tangential distortions and solve, in the sense of least squares, for the relevant parameters. Finally, the two channels can after rectification, reach an ideal configuration similar to that of the canonical stereo camera.

- Chapter 5 specifies how the problem of the determination of features in transitional areas is an ill-posed problem, and discusses how to deal with this. Afterwards, I propose two approaches to detect transitional areas: one based on intensity and another based on the signal gradient. Correspondingly, two methods of subpixel positioning for a pair of feature pixel points are adopted, which include the intensity mean and reconstructing the figure of an edge. I employ synthetic channel images and real images captured by eCley to compare the results, and analyze the illumination sensitivity, the object distance sensitivity, and the interval of object movement.

- Chapter 6 demonstrates the essential preprocessing for real-time channel images after giving an overview of real-time subpixel distance measurement for objects. Then I use the coarse-to-fine strategy to obtain the subpixel position of a pixel pair, i.e. a fine interpolation needed after using the Canny edge detector. The difference between the subpixel positions of a pair of feature pixels is just the disparity of the pixel pair, which yields the subpixel distances of objects by triangular geometry. I perform the comparison experiments and analyze the factor of distance measurement as well.

- Chapter 7 concludes with all the key points of my whole work and the real application, which is followed by outlining avenues of future work on improving the precision and efficiency of the computation, on reducing the effects of the environment, and on obtaining dense disparity.

Chapter 2

Stereo Vision in the Artificial Compound Eye

2.1 Conventional Stereo Vision

Stereo vision enables non-living systems, such as a robot, to perceive the depth information of the real world in a similar way as the eyes of human beings do. Although a variety of algorithms have been proposed in last decades, generally, dense stereo techniques are divided into two categories: local approaches and global approaches. Certainly, all of these approaches base on the canonical stereo principle, which will be explained in following section.

2.1.1 3D Measurement from a Canonical Stereo Vision

One of the advantages of the stereo approach to 3D reconstruction and object recognition is that the geometrical relationship between left camera and right camera is already known due to the fixed configuration of most stereo systems. Usually the geometrical relationship can be inferred easily from the ideal standard (called canonical) system.

A standard binocular stereo camera system is illustrated in Figure 2.1. Generally, the two cameras are mounted such that their optical axes are coplanar and aligned in parallel. The span between the optical centers of the left (O_l) and right (O_r) cameras is called the baseline B. The middle point of baseline B is set up as the origin $O(0,0,0)$ in the Camera Coordinate System (CCS), i.e. the axis Z of coordinate system is orthogonal to the connecting line of the two optical centers of the cameras. The two cameras have the same focal length f from the image planes to optical centers respectively. The interaction of the optical axis of the left camera with the image plane is the principe point P_l, likewise P_r is principle point of the right camera. The image point $m_l(x_l, y_l)$ on the left image plate and $m_r(x_r, y_r)$ on the right image plane are projective

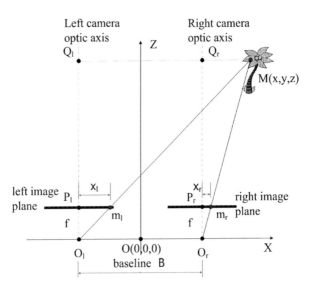

Figure 2.1: Stereo vision principle. Standard (canonical) system of two cameras with focal lengths f, and a base-line B.

respectively from the same 3D objective point M(x, y, z) (x,y and z are 3D space coordinate in CCS) in the scene; here the tree. Considering the similar triangles $\Delta P_l m_l O_l$ and $\Delta Q_l M O_l$, where Q_l denotes the projective point with respect to the 3D point M(x, y, z) on the principle axis of left camera, we obtain the equation

$$\frac{f}{z} = \frac{x_l}{\frac{B}{2} + x}. \tag{2.1}$$

For the right camera, the similar triangles $\Delta P_r m_r O_r$ and $\Delta Q_r M O_r$ are used, Q_l is the projective point with respect to the 3D point M(x, y, z) on the principle axis of right camera. The relationship can be described as

$$\frac{x_r}{x - \frac{B}{2}} = \frac{f}{z}. \tag{2.2}$$

By the combination of (2.1) and (2.2), the depth z can be represented as

$$z = \frac{Bf}{x_l - x_r}, \tag{2.3}$$

where $x_l - x_r$ is referred to as disparity, which is not only the most important argument in stereo vision systems but also the ultimate goal to stereo matching algorithms. For obtaining the x coordinate of the 3D object, we substitute (2.3) into (2.1) and derive the distance x as

$$x = \frac{B(x_l + x_r)}{2(x_l - x_r)}.$$ (2.4)

As well, we can utilize the triangle similarities in the Y-Z plate to attain the following relationship:

$$\frac{y_l}{f} = \frac{y_r}{f} = \frac{y}{z}.$$ (2.5)

Finally, based on (2.3) and (2.5), the y coordinate of the scene point M can be calculated by

$$y = \frac{By_l}{x_l - x_r}.$$ (2.6)

Thus, the coordinates (x, y, z) of the 3D object point M can be exactly derived from the 2-D coordinates (x_l, y_l) and (x_r, y_r) with respect to the known projective points m_l and m_r respectively. In other words, through the positions of two image points in stereo vision system, we can determine the position of the corresponding 3D object point using the following equation:

$$\begin{pmatrix} x \\ y \\ z \end{pmatrix} = \begin{pmatrix} \frac{B(x_l+x_r)}{2(x_l-x_r)} \\ \frac{By_l}{x_l-x_r} \\ \frac{Bf}{x_l-x_r} \end{pmatrix} = \frac{B}{x_l - x_r} \begin{pmatrix} \frac{x_l+x_r}{2} \\ y_l \\ f \end{pmatrix}.$$ (2.7)

2.1.2   Related Work in Stereo Vision

The main challenge of stereo vision is the reconstruction of 3D information of a scene from two images taken from distinct viewpoints. The significant tasks of stereo vision are calibration, correspondence and 3D reconstruction. Generally, we assume the two cameras rigidly meet to a specific set up structure, the so called canonical stereo system [Cyganek & Siebert, 2009]. If pixel correspondences between both images can be found, the depth information can be attained easily and the 3D reconstruction can be done. Based on the canonical stereo system, computation of disparity (defined as the horizontal displacement of corresponding pixels) becomes the main goal of stereo matching algorithms. More intuitively, all the disparities of pixels are exactly represented by a disparity map with the same size as the stereo images, which contains the disparity for each pixel as an intensity value. In the ideal case, the disparity of two corresponding points can be determined uniquely since one scene point is projected to two image points with the same identifiable features. Actually, there are many factors such as noise which affect the imagery of cameras in the real world. So, the image point in the left image is not completely matching with the corresponding image point in the right image. That means, both points lost some common features and some dissimilarities arise. Therefore, many troublesome problems need to be overcome in stereo matching algorithms. The most common problems stereo matching algorithms have to face are occluded areas, reflections in the image, textureless areas or periodically textured areas and very thin objects [Humenberger *et al.*, 2010].

A notable number of approaches have been proposed in the last years to solve the problem of stereo correspondence, and most of them determine stereo disparities by exploiting different constraints. All of these methods attempt to match pixels in one image with their corresponding pixels in the other image. There are two main categories of stereo matching algorithms: local and global algorithms. On the whole, local algorithms involve constraints on a small number of pixels surrounding a anchor pixel. Similarly, global algorithms based on global constraints loosely involve constraints on scanning lines or on the entire image [Brown *et al.*, 2003]. Local algorithms are traditionally characterized by efficient and simple approaches. Although being capable of achieving real-time frame rate performance, they typically fail on low-textured areas as well as along depth borders and over occluded regions. Global algorithms exploit nonlocal constraints, and this additional support reduces the sensitivity to local regions in the image that fail to match, due to occlusion, uniform texture, etc. However, the use of these constraints makes the computational complexity of global matching significantly greater than that of local matching.

2.1.2.1 Local matching methods

On the one hand, local matching methods can utilize the features of an image, such as corners, curves or edges, to find the proper corresponding parts in another image, and reach the stereo matching. On the other hand, the block of areas around each pixel can also be used to attain matching pixel pairs all over the stereo image pair.

Feature-based algorithms appear rather robust to the regions of uniform texture and depth discontinuities in images, since they limit the regions of support to specific reliable features in the images. Of course, this also limits the density of points for which depth may be estimated, i.e. only a sparse disparity map is generated. Due to the efficiency of feature-based algorithms, many researchers paid their attention to this direction of work. Schmid and Zisserman proposed an algorithm to automatically match individual line segments and curves between images [Schmid & Zisserman, 2000]. The algorithm uses both photometric information and the stereo geometric relationship. The homography facilitates the computation of a neighbourhood cross-correlation based matching scores for putative line or curve correspondences. Venkateswar and Chellappa have proposed a hierarchical feature-matching algorithm exploiting four types of features: lines, vertices, edges, and edge-rings (i.e. surfaces) [Venkateswar & Chellappa, 1995]. Matching begins at the highest level of the hierarchy (surfaces) and proceeds to the lowest (lines). It allows coarse, reliable features to provide support for matching finer, less reliable features, and it reduces the computational complexity of matching by reducing the search space for finer levels of features. Driven by the need for dense depth maps for a variety of applications and also due to improvements in efficient and robust block matching methods, the interest in feature-based methods has declined in the recent decade [Brown *et al.*, 2003].

According to Scharstein and Szeliski's research [Scharstein & Szeliski, 2002], basically, the area-based algorithms, also called block matching, fulfill the stereo matching trough the following four steps: Firstly, a preprocessing should be performed to filter noise, balance brightness or improve contrast of the image. Secondly, the matching cost for each pixel at each disparity level in a certain range is computed and cost aggregation is done in support windows. Then, the correct correspondence, i.e. the optimal match, is determined by comparing the dissimilarity or similarity. Finally, the improved and smooth disparity map is obtained by refinement processing.

Area-based algorithms try to match each pixel independently to their correspondences without the effect of the image content and calculate the disparity for each pixel in the image. Thus, the resulting disparity map can be very dense. Usually, area-based algorithms employ different similarity or dissimilarity measures to attain exclusive performance respectively.

Normalized cross correlation (NCC) and the Zero-mean Normalized Cross-Correlation (ZNCC) are the standard statistical method for determining similarity. Its normalization, both in the mean and the variance, makes it relatively insensitive to radiometric amplitude and offset. So, these two methods are typically employed when robustness with regard to photometric variations is required. On the other hand, a common class of dissimilarity measures is derived from the Lp norm. The two most popular Lp norm-based dissimilarity functions are the Sum of Squared Differences (SSD) and the Sum of Absolute Differences (SAD). In the following, popular match cost calculation approaches are shown, based on similarity and dissimilarity metrics for the pixel (x, y) in both, the reference image I_1 and in the corresponding image I_2 are shown. The size of the support window is $m \times n$, the disparity is represented as d, and $C_{...}$ denotes the calculated match cost depending on method ...

$$C_{SAD} = \sum_{i=m} \sum_{j=n} |I_1(x+i, y+j) - I_2(x+i+d, y+j)| \qquad (2.8)$$

$$C_{SSD} = \sum_{i=m} \sum_{j=n} (I_1(x+i, y+j) - I_2(x+i+d, y+j))^2 \qquad (2.9)$$

$$C_{NCC} = \frac{\sum_{i=m} \sum_{j=n} I_1(x+i, y+j) I_2(x+i+d, y+j)}{\sqrt{\sum_{i=m} \sum_{j=n} I_1(x+i, y+j)^2 \sum_{i=m} \sum_{j=n} I_2(x+i+d, y+j)^2}} \qquad (2.10)$$

13

$$C_{ZNCC} = \frac{\sum_{i=m}\sum_{j=n}(I_1(x+i,y+j)-\bar{I_1})I_2((x+i+d,y+j)-\bar{I_2})}{\sqrt{\sum_{i=m}\sum_{j=n}(I_1(x+i,y+j)-\bar{I_1})^2\sum_{i=m}\sum_{j=n}(I_2(x+i+d,y+j)-\bar{I_2})^2}}$$

(2.11)

where

$$\bar{I} = \frac{1}{mn}\sum_{i=m}\sum_{j=n}(I(x+i,y+j))$$

Because area-based matching algorithms can very efficiently compute the disparity map, these are suited for real-time application [Hirschmüller et al., 2002]. The quality of the results is not yet comparable with global methods. As we know, the aggregation windows and manner are significant factors in local matching algorithms. For increasing the accuracy of disparity estimations, a number of area-based algorithms focus on the central problem that is how to determine the size and shape of the aggregation window. More accurately, a window must be large enough to cover sufficient intensity variation while at the same time it must be small enough to avoid crossing depth discontinuities for a reliable estimation.

In the early stage, the traditional area-based method is the simplest fixed window or box filter method, in which a fixed (square) window is assigned to each anchor pixel as its support region. Although the method has the lowest complexity and highest efficiency, it easily blurs object boundaries [Zhang et al., 2011]. Aiming at the problem of low accuracy along depth borders, a variable support window or multiple windows to compute the local matching cost rather than using a fixed squared window was adopted [Veksler, 2003]. This method selects one window or a combination of several windows from a number of candidates with respect to different anchor points such as the points at edges, discontinuous and textureless areas, where they produce a small matching cost. They attain much better results than the fixed window method. However, due to the fact that the shape of the support windows is limited to a small number of candidates, it is difficult to preserve delicate structures in disparity maps. The adaptive shape methods [Lu et al., Jan. 2008] construct shape-adaptive support regions by approximating the local image structures. Unlike the multiple windows methods, the shape of support regions in adaptive shape methods is more flexible and is not restricted to be rectangles. For achieving high matching accuracy, Zhang et al. [Zhang et al., 2009a] constructed an arbitrarily shaped support regions in a separable way and accelerated cost aggregation by a so-called orthogonal integral images technique. That processing brings the benefit that it is two orders of magnitude faster than the adaptive weight methods.

Among all the local area-based algorithms, the ones based on adaptive weight cost aggregation give the finest disparity maps. This method uses a large size support window and assigns a support weight to each pixel in the window. The weight is calculated based on Gestalt principles [Yoon & Kweon, 2005], which state that the grouping of pixels should be based on their intensity difference and spatial distances to the anchor pixel. In other words, the pixels with similar intensity and small distance to the anchor pixel are more likely to have the same depth as the anchor pixel. These adaptive weight methods are able to yield very accurate disparity maps with the object boundaries preserved well. To a large extent, they are comparable with global methods for high quality disparities. But they are computationally quite expensive and require a huge amount of memory since the window must be big enough for the aggregation to be effective [Tombari et al., 2008][Lu et al., Jan. 2008]. Nevertheless, due to parallelism, these methods can be speeded up if ported to programmable graphics hardware [Kowalczuk et al., 2013][Zhang et al., 2011], so they can broadly be applied to many real-time conditions.

2.1.2.2 Global matching methods

Global methods exploit an optimization process for all cost values to determine the best stereo matching. They reduce sensitivity to local regions in the image that fail to match, due to occlusion, uniform texture, etc. This is because obtaining all local cost values and satisfying other constraints can be reached simultaneously as the disparity values are found, which fit best into this optimization task. Most global methods attempt to minimize an energy functional computed on the whole image area by means of a Pairwise Markov Random Field model (MRF). Since this task turns out to be a NP-hard problem, approximate but efficient strategies such as Graph Cuts (GC) and Belief Propagation (BP) have been proposed. The energy functional for stereo matching can be stated as [Scharstein & Szeliski, 2002]

$$E(d) = E_{data}(d) + \lambda E_{smooth}(d). \tag{2.12}$$

The data term, $E_{data}(d)$, measures how well the disparity function d agrees with the input image pair. Using the disparity space formulation,

$$E_{data}(d) = \sum_{(x,y)} C(x, y, d(x, y)), \tag{2.13}$$

where C denotes the matching cost value. The term $E_{smooth}(d)$ is introduced to enforce a smoothness of the solution, i.e. an additional constraint on the resulting disparity map. Sometimes it is additionally related with a function of image intensity, but usually it is a function of disparities restricted to only measuring the differences between neighboring pixels disparities,

$$E_{smooth}(d) = \sum_{(x,y)} f(d(x, y) - d(x + 1, y)) + f(d(x, y) - d(x, y + 1)), \tag{2.14}$$

15

where $f(\cdot)$ denotes a monotonically increasing function on its argument of disparity.

According to diverse global optimization approaches, global matching methods can briefly be categorized to dynamic programming, graph cut, diffusion, and belief propagation.

The main idea of dynamic programming lies in division of the 2D search problem into a series of separate 1D search problems on each pair of epipolar lines. The ordering constraint, which means that pixels in the reference image have the same order as their correspondences in the matching image, specifies the possible predecessors of all matches. The path with the lowest matching costs is chosen recursively. This leads to a path through the possible matches that implies a left/right consistency check. Some variant approaches that focused on the dense scanning line optimization problem work by computing the minimum-cost path through the matrix of all pairwise matching costs between two corresponding scanning lines [Birchfield & Tomasi, 1998] [Bobick & Intille, 1999]. Despite the resulting disparity maps usually suffer from horizontal streaks caused by only considering horizontal smoothness constraints, the running efficiency makes it broadly applicable to real-time environments.

The approach called graph cuts, which exploits the two-dimensional coherence constraints available, avoids the drawback of the dynamic programming methods. This method assumes the solution of the stereo problem as the computation of a maximum flow in graphs. This graph has two special vertices, the source and the sink. Those nodes are connected by weighted edges. Each node represents a pixel at a disparity level and is associated with the according matching costs. Each edge has an associated flow capacity that is defined as a function of the costs of the node it connects [Kolmogorov, 2004]. This capacity defines the amount of flow that can be sent from source to sink. The maximum flow is comparable to the optimal path along a scanning line in dynamic programming, with the difference that it is consistent in three dimensions. The computation of the maximum flow is extensive, so it cannot be used for real-time applications [Tappen & Freeman, 2003].

Belief propagation is another global optimization approach, which formulates the stereo matching problem in a probabilistic way by means of Markov random fields. From the maximum a posteriori estimation is obtained by applying a Bayesian belief propagation (BP) algorithm. BP performs a kind of message passing, where the message is meant as a probability that a receiver (a node in MRF) should exist disparity, which is congruent with all information already passed to it by a sender. The nodes are divided into high-confidence and low-confidence ones. The entropy of a message from high-confidence nodes to low-confidence nodes is smaller than in the opposite direction. At every iteration, each node sends its belief value that is the sum of the matching costs and the received belief values to all four connected nodes. The new belief value is the sum of the actual and the received value and is saved for each direction separately. This is done for each disparity level. Finally, the best match is the one with the highest belief values defined by a sum over all four directions [Klaus *et al.*, 2006][Yang *et al.*, 2009].

At present, most global approaches exploiting segmentation provide the most accurate results on the Middlebury dataset [Shi *et al.*, 2012]. However, global approaches are computational expensive and hence currently not suitable for real-time application.

2.1.3 Analysis of Depth Resolution

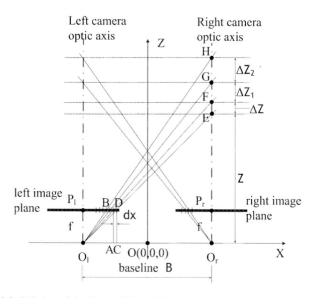

Figure 2.2: Relation of depth sensitivity with respect to camera resolution and horopter

From the formula for 3D distances measurement (2.7), a major limitation of stereo reconstruction can be inferred, i.e. the depth resolution decreases rapidly with increasing distance between the objects and cameras. Figure 2.2 explains not only this phenomenon but also the dependence of the depth accuracy versus camera resolution and distance to the observed scene. Each line (such as the line with E) represents a plane of constant disparity with respect to a fixed depth from camera in integer pixels. We define the horopter as the range from the minimum disparity to the maximum one, which represents a 3D volume covered by the search range of the stereo algorithm. Each disparity limit (the number of disparities between the maximum and minimum) sets a different horopter at which depth can be detected. As shown in Figure 2.2, for instance, the minimum and maximum disparities are the line with E and the line with G respectively. So, the horopter with the limit of three pixels is the range from the line with E to the line with G; while the range between the line with F and the line with H become another horopter with the limit of three pixels at the depth $(Z + \Delta Z)$. Outside of this range, depth will not be found and will represent a hole in the depth map where depth is not known. Observing the triangle $\triangle AO_l B$ is similar to $\triangle O_r O_l F$, we get the following relation:

$$\frac{\overline{AO_l}}{f} = \frac{B}{Z + \Delta Z}. \tag{2.15}$$

Likewise, $\triangle CO_l D$ is similar to $\triangle O_r O_l E$, and the relation is

$$\frac{\overline{CO_l}}{f} = \frac{B}{Z} \tag{2.16}$$

The known size of pixel is defined as $\Delta x = \overline{CO_l} - \overline{AO_l}$, thus, the combination of (2.15) and (2.16) becomes

$$\Delta Z = \frac{\Delta x Z^2}{Bf - \Delta x Z} \tag{2.17}$$

Assuming now that Bf/Z is much larger than the pixel resolution Δx, we obtain the following approximation:

$$\Delta Z = \frac{\Delta x Z^2}{Bf} \tag{2.18}$$

Due to the baseline B and focal length f are a constant values, obviously, the depth resolution decreases with the square of distance Z^2. This means that if it is necessary to measure the absolute position of real objects with an a priori assumed accuracy, then the parameters of the stereo setup must be chosen in such a way that R would be at least an order of magnitude less than the assumed measurement accuracy [Cyganek & Siebert, 2009]. As we can see in Figure 2.2, there are different uncertainty values $Z, \Delta Z, \Delta Z_1$ at different depths $Z, Z + \Delta Z, Z + \Delta Z + \Delta Z_1$.

In order to analyze the depth resolution in detail, I take an example of a real stereo vision system with baseline $200\ mm$, focal length $25\ mm$ and pixel's size $3.2\ \mu m$. Figure 2.3 shows the results of the horopter with the horopter set to 5–20 and a disparity range of 15, giving a measurable range of around 40–$310\ m$. The depth resolution is shown in Figure 2.4. When the measurement accuracy is required to be $1m$, the measurement distance should be kept lower than $40\ m$. If the object lies in the position of $100\ m$, the uncertainty of measurement will reach a quite large number of $6.4\ m$.

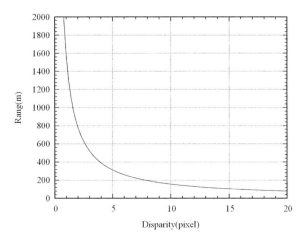

Figure 2.3: Variation of depth with respect to disparities

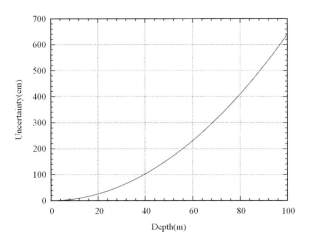

Figure 2.4: Uncertainty values with respect to depth

2.2 Current Stereo Vision in an Artificial Compound Eye

Since artificial compound eyes provide useful properties, the application field of digital imaging system is extending more and more. At present, intensive studies are focused on color imaging, multispectral imaging, and fingerprint capturing [Horisaki *et al.*, 2008].

As for the application of stereo vision to the artificial compound eye, some researchers are still dedicated to develop some new approaches to obtain 3D information and reconstruct high-resolution images. R. Horisaki etc. [Horisaki *et al.*, 2007] proposed an approach of three-dimensional information acquisition using a compound imaging system based on the modified pixel rearrangement method and the estimation of the object distance. This method utilizes multiple images observed from different viewpoints, which are captured by a kind of artificial compound eye so called TOMBO (thin observation module by bound optics), to determine depth information and reconstruct high-resolution images from multiple low-resolution images rather than only to improve the resolution of the reconstructed image. First, the method projects the pixels on the related channel (unit) images to the virtual image plane and set up the corresponding pixel values on the virtual image by a weighted average. Afterward, the sum of squared difference (SSD) between the base image (projected image) and all the reference images (back-projected images) is calculated as objective function. Finally, the distance map is estimated by minimizing the SSD and reconstructing the 3D high-resolution image. Although the correct distances of objects were successfully estimated, the SSD is sensitive to the variation in brightness of the channel images. R. Horisaki etc. [Horisaki *et al.*, 2009] introduced a normalized cross-correlation to replace the SSD due to its robustness against brightness variations between the images and employed the iterative back-projection algorithm to implement an image reconstructed scheme considering color shift, brightness variation, and defocus. Evidently, most methods of 3D reconstruction in artificial compound eyes are based on TOMBO and multiple channel images. Unfortunately, there is no approach developed exclusively for eCley type of facet camera. A fact is that the determination of 3D information in eCley becomes more and more desirable with the improvement of other outstanding functions of eCley and comprehensive application.

2.3 Limitations of General Stereo Vision for eCley

Generally, the outstanding properties of the eCley are the miniature volume and short focal length. However, these exclusive properties make some conventional approaches unsuitable for them anymore. In this case, the miniature imaging systems can not only implement the common functions of classical single aperture cameras but also achieve some exclusively other properties such as the use as a micro stereo channel array. Thus, it is important to discuss the limitations of conventional approaches for the eCley, in order to find the solution space where a compound optical array is appropriate. An instance of eCley has a attracting features of 17 × 13 channels with the same diameter, which keep rigidly horizontal and vertical with the scanlines of the sensor, namely the centers of each row or column always stay in the same line [Brueckner et al., 2010]. Thus, the adjacent two channels naturally comprise a micro stereo camera. Is this micro stereo system able to use conventional stereo matching methods to get satisfying disparities? The following limitation-discussions involving the short baseline, short focal length and size of imagery plate will answer it.

In the context of stereo vision, the size of baseline is considered as a tradeoff between precision and correctness in stereo matching. For a longer baseline, there is certainly a longer disparity range in the stereo vision system, in which stereo searching and matching become rather difficult. On the other hand, the occlusion arises more frequently and a greater possibility of false matches is encountered. A short baseline between a pair of stereo images makes the distances of objects less precise due to the narrow triangulation geometry shown in Figure 2.2.

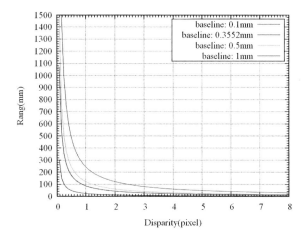

Figure 2.5: Range with respect to disparity and different micro baselines

Usually, the baseline of eCley is quite short at the order of magnitude of a few hundreds micron. For instance, assuming one of the eCley with the channel pitch (baseline between two adjacent channels) 355.2 μm, focal length 778 μm and pixel pitch (size of pixel) 3.2 μm, we can obtain the relation of the depth range with respect to disparity and different baselines shown as Figure 2.5. Obviously the measurement range of this stereo system indicated by the blue curve

21

is quite narrow around $100 \ mm$ even if the maximum disparity is increased over 8 i.e. enlarge the horopter to 8, 16, or 64. Whether the baseline can be increased to two or more times, e.g. to $1 \ mm$, the measurement range can be enlarged but still it is around $220 \ mm$. That is too small to be applied to the real environment broadly.

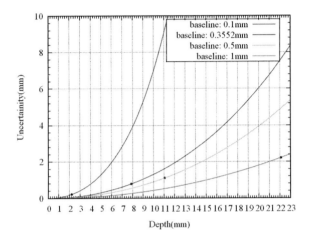

Figure 2.6: Uncertainty with respect to depth and different micro baseline

On the other hand, the results of minimization of eCley form a challenge to the precise of stereo measurement. As shown in Figure 2.6, the accuracy over depth with different baselines still is consistent with common stereo vision systems. I.e. the uncertainty of depth evidently increases with the distance of object to the camera, but their uncertainties become more severe than that of common stereo systems. The black points on each curve indicate the uncertainty of 10% with respect to depth. For the assumed eCley, the black point on the the blue curve shows that the system has high precision (uncertainty less than 10%) only under $8 \ mm$, while the depth measured out of the range is rather rough and uncertain.

According to (2.3) and (2.17), the properties of the focal length vary similarly to that of the baseline in stereo vision of eCley. In that case, the conventional stereo vision only results in an extremely small and narrow measurement range lower than one hundreds millimeters. On the other hand, the uncertainty of the stereo vision measurement becomes quite severe at common distances of objects for the eCley. More accurately, the confident interval limits in the rather narrow area in a statistical perspective i.e. the acceptable precision of measurement is only around a quite near range such as decades of millimeters.

Usually, the conventional methods do not detect the desired features easily in the case of images with rather small sizes. Even if images are enlarged, the results are not improved, because the noise is also multiplied and the information content is not increased. As the (2.3) describes the relationship between the disparity and the distance of the object, when B, the baseline of the stereo camera is known, the depth information that represents the distance between the object and the camera can be derived from the disparity of the stereo pair [Cyganek & Siebert, 2009]. Assuming one eCley channel has a diameter of 79 pixels and the offset between channels is 17 pixels, then the maximal and minimal distances of the objects should be between $86\ mm$ and $1.4\ mm$. In such a small range, if using the stereo matching, the edge size of objects detected should not be more than 48 mm, which extremely confines the application of the eCley. Additionally, in the case of placing an object over the maximal distance, i.e. with a disparity smaller than one pixel, the disparity will not appear and will not be detected in the images whatever the size of the objects. Certainly, the limitation can also be seen in Figure 2.7. The place where the chessboard lies in is near maximal distance of the disparity. However, the valid matching area (blue area) is only approximately two third of the area of the channel, i.e. about 62 pixels for the disparity. It is conceivable that such a small disparity range will hardly make the conventional stereo matching algorithms work effectively.

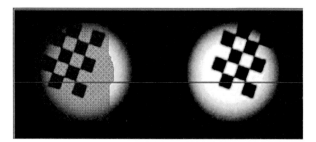

Figure 2.7: The valid stereo area of two microlens stereo channels after calibration

Therefore, the common stereo methods are not suitable for the eCley, and new techniques or methods have to be developed to find a reliable way to obtain the depth information. For instance, the intensity properties of edges of the objects can be considered for this task.

Chapter 3

Subpixel Stereo Vision Based on Variation of Intensities

Aiming at the narrow disparity range and low precision, subpixel processing is employed for the implementation of stereo vision for the eCley. But this subpixel process is not similar to that in conversional stereo matching algorithms, which utilizes post processing to improve the outcomes at pixel level. It uses a sequence of images captured from a moving test model to calibrate the subpixel distance between each pixel pair, and then really detects the difference of the correspondences from objects, to attain the depth information.

3.1 Subpixel Resolution

Since the effect of minimization of the eCley, the resolution of each channel image is not high and even quite low as the images captured by common camera. Therefore, the processing results from these channel images won't be of high precision unless getting supports from additional information or special processing methods. Subpixel resolution exploits more details of a scene in an image compared to the image described as pixels. Usually this goal of improving precision can be reached by various algorithms rather than hardware, which obviously saves resources and cost and facilitates itself to apply to broader fields.

3.1.1 Subpixel Stereo Matching

In most stereo matching algorithms, subpixel estimation generally is used as post processing for high precision. As we know, disparity estimation essentially computes the distance between the positions of the stereo matching points in the corresponding images. Due to the discretization of sampling in the real scenes, the minimal accessible unit of images is one pixel. Therefore, the disparity based on the integer grid of pixels can only have integer values.

Although, in most applications such as robot navigation or object recognition and tracking, pixel accuracy may be perfectly adequate, for image-based rendering and object measurement under low image resolution, such quantized maps lead to excessive errors and very unappealing results. For increasing the resolution of disparity, deducing the size of pixels is the ideal way to obtain high precision and to get satisfying processing speed. But the cost of hardware and the technique of implementation are main restrictions at present. Alternatively, we can use software methods to reach the goal e.g. using interpolation. Actually, many algorithms apply a subpixel refinement stage following to the initial discrete correspondence stage. This way we can attain a more precise position of a minimum of the matching measure (i.e. the cost function), which does not need to fall exactly at the integer pixel position due to continuous support.

In general, these subpixel matching algorithms can be classified into interpolation-based (fitting), phase correlation-based, and optimum parameter estimation-based. The most popular and intuitive subpixel methods are implemented by interpolation among the discrete sample points. Since the optimal matching should be at the minimal or maximal value, the three values of a matching measure within the neighborhood of a interest point is usually fitted to a third-order curve, a parabola [Yang *et al.*, 2009]. Then, the position of a minimum of this parabola is found, which indicates a new disparity value with subpixel resolution. Certainly, a larger number of points can be used to interpolate higher order polynomials and possibly to increase precision further. Nevertheless, processing a larger number of points extremely increases computational cost. Considering the trade-off between precision and efficiency, in practice fitting a parabola is the most efficient method in terms of accuracy achieved versus computational effort [Mühlmann *et al.*, 2002]. Another interpolation method is intensity interpolation. It seeks a peak position of the similarity in high-spatial-resolution images obtained by image interpolation [Szeliski & Scharstein, 2002]. Generally, this method requires much computation time and memory space. But some improved methods are proposed to remedy these defects. They use a continuous function or hierarchical coarse-to-fine measures to interpolate images [Shimizu & Okutomi, 2005]. Nehab et al. demonstrated the strategy of symmetric subpixel refinement for accurate correspondences [Nehab *et al.*, 2005]. This method uses the symmetry to refine the coordinates of corresponding points in both reference and matching images, further retains the detail and avoids the systematic error that stems from the asymmetry between the reference image described as discrete integer grid pixels and the matching image described with subpixel resolution. Furthermore, an improved variant of the symmetric subpixel refinement even overcomes the phenomenon of "pixel blocking" [Nehab *et al.*, 2005][Shimizu & Okutomi, 2005].

The seeking correspondence problem can be seen as a mapping between coordinate systems of two images. Further, this problem can be reduced to an optimum parameter estimation problem for the mapping transformation, which defines objective functions and requires nonlinear optimization techniques to reach the optimum. Lucas et al. very early proposed an iterative image registration technique for subpixel stereo correspondence [Lucas & Kanade, 1981], which uses the spatial intensity gradient information to direct the search for the position that yields the best match. But the algorithms convergence and accuracy to a large extent depends on the accuracy of the initial estimates for depth and on parameter vector. A regularization framework for iteratively refining subpixel disparities is proposed by Yu and Xu [Yu & Xu, 2009], which selects normalized cross-correlation as the matching metric, and performs disparity refinement based on correlation gradients instead of intensity gradient-based ones. To reduce the effect of image noise, Nefian introduced a Bayesian formulation to replace the cost function and then optimally estimate the parameters of the noise model [Nefian et al., 2009], which reduces significantly the pixel locking effect of earlier methods as well.

Phase correlation adopts the good properties of the Fourier transformation, i.e. the transformation such as shift, rotation and scale in space domain have their corresponding properties in the Fourier domain respectively. Extensions of phase correlation can reach a subpixel matching by detecting the phase differences in the Fourier domain which correspond to a shift in spatial domain. An analytic expressions for the phase correlation of downsampled images has been proposed by Foroosh et al. [Foroosh et al., 2002], which provides a closed-form solution to subpixel translation estimation, and are used for detailed error analysis as well. Foroosh et al. put forward that the discrete phase difference of two shifted images is a two-dimensional sawtooth signal. Based on this prerequisite, the phase difference can be determined directly for subpixel accuracy by the number of periods of the phase difference along each frequency axis [Foroosh & Balci, 2004]. Therefore the improved method greatly decreases the computational cost and increases robustness to noise. A new enhanced correlation-based similarity with invariability in photometric distortions and subpixel accuracy is introduced by Psarakis and Evangelidis [Psarakis & Evangelidis, 2005]. Clearly, the optimum value of the displacement is the closed-form solution derived from the optimum solution of the parameter model. In general, most subpixel approaches are used as a post processing, i.e. based on the initial disparity obtained at pixel. So, the results depend heavily on the first step. They are estimated values but not real values. Even though some algorithms derive the subpixel disparities from the intensity images, they are not suitable for the eCley due to the limitation of minimized structure of the eCley (see chapter 2 for more details).

3.1.2 Subpixel Edge Detection

The other class of subpixel approaches broadly arises in the context of edge detection. According to the formation of digital images, the digital raster limits the resolution of an image to one

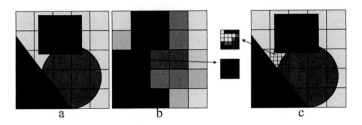

a b c

Figure 3.1: Mixed pixel problem and subdivided processing

pixel. Because one pixel only has an intensity value when it's sampled and quantized, the image shape may be changed or unambiguously registered compared to the original scene, especially in mixed pixels. In the case of mixed pixels, the intensity of pixels is a mixed value from the neighboring objects, i.e. it is affected by the different regions in its neighborhood. Although we can generally increase the image resolution to improve the quality of image processing, this problem is not eliminated essentially. For a better explanation of processing mixed pixels, Figure 3.1 shows the overlap region in mixed pixel and the processing by increasing the resolution. As the left image (a) shows an original scene covered by a pixel grid, the objects (triangle, circle and rectangle) are not totally aligned with pixels. Especially, the pixel at row 3 and column 2 overlaps on three object regions which commonly affect the intensity of the mixed pixel. The second image (b) is the output after digitalization, which uses a uniform intensity value to set the mixed pixel. From the right image (c), we can see that the original scene is subdivided through increasing image resolution, more accurately, the pixel intensity increased the quintuple image resolution, which are shown in two close-ups.

Generally, these mixed pixels compose common edge points and all of them become object edges. Due to this, pixels affected by objects and background together. These neighborhoods of edges can also be seen as transitional areas where objects change to background and vice versa. Conventional algorithms of edge detection determine the results at pixel precision, actually in most cases, real edges lie at some place inside a pixel. Thus, subpixel techniques usually are used for high precision edge position computation. Notice, that the image rasterization not only reduces the edge information of scenes but also is an irreversible process. We only get the approximate edge positions. At present, a variety of subpixel edge detection algorithms are proposed to estimate reasonable edge positions and they can be classified into reconstructive methods, moment-based methods and curve-based methods [Fabijaska, 2012]. Reconstructive methods are based on the intensity distribution of pixels in transitional areas and derive a continuous probability function from the intensity distribution for estimating subpixel edges. A well implemented approach based on the intensity reconstruction uses a second order polynomial to approximate image intensity at the edge. The expected subpixel position of the edges can be indicated by a point at the reconstructed curve where the intensity equals to the mean of the object and background [Xu, 2009]. When using the first derivative of image intensity, the edges are comprised of points where the deviation reaches a maximum in transitional areas. The subpixel positions of the edges are derived from sampling in the gradient neighborhood and reconstructing a gradient function in the intensity-changing transitional areas. Certainly, the subpixel point should be located at the maximum value of the gradient function [Fabijanska & Sankowski, 2010][Liu *et al.*, 2004]. In many applications, the cubic and Gaussian functions can well fit the gradient curves in transitional areas. Since the edges are also defined as the zero-crossing position in second derivative of image intensity, a method of edge detection is based on finding zero crossings of the second derivative of a signal (so-called zero crossing operator). In other words, a continuous second derivative function can be reconstructed in each zero-crossing point neighborhood which is obtained by operators such as Laplacian of Gaussian (LoG) [Perez *et al.*, 2005]. The edge with subpixel accuracy is located at the point where the zero-crossing of the reconstructed second derivative function is located [MacVicar-Whelan & Binford, 1991]. In summary, reconstructive methods only use the few pixels in transitional areas to determine subpixel positions. So, they can run fast and efficiently. But the methods will be quite sensitive to noise regarding to the normal number of points. The main reconstructive methods are illustrated in Figure 3.2. By the digital sampling and discretization, a real sharp edge (a)) becomes a transitional area in the raster image where the intensities of pixels change gradually from the intensity of object to that of the background. Using the common methods of edge detection, the precision at most reaches one pixel because one pixel is the minimal unit of an image. Through the function reconstructions of the intensity distribution, the first derivative of intensity and second derivative of intensity, the edges are located at accurate subpixel position in transitional areas such as e) and f). Obviously, the determined subpixel positions are quite near to the real one.

Moment-based methods detect the subpixel positions of edges based on the principle of invariant moments, i.e. the moments of objects are unchanged before and after imagery. On the whole, these methods are clearly divided into these using geometrical moments and those with orthogonal moments. Further, Gray moments based on pixel intensities and spatial moments which utilize spatial information about the neighborhood are distinguished in geometrical moments [Tabatabai & Mitchell, 1984][Lyvers *et al.*, 1989]. Gray moment-based methods assume the Gray moments in the real edge distribution is an invariant like in the ideal step edge model. This edge detector based on Gray moments employs an unit circle comprised of a sequence of pixels as input data and two-dimensional ideal step edge model that is seen as the combination of a sequence of pixels with intensity value h_1 and the other pixels with intensity value

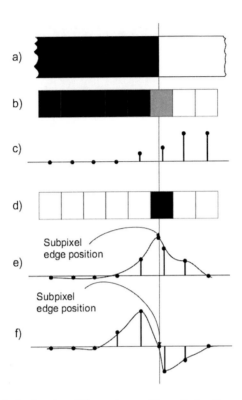

Figure 3.2: Subpixel edge based on different classes of intensity function reconstruction: a) real scene, b) image of the real scene, c) intensity profile of the scene, d) edge position at pixel level, e) subpixel edge position result from the first derivative, f) subpixel edge position result from the second derivative

h_2. All the parameters: edge position ρ, edge direction θ, and both intensity values h_1 and h_2 are derived from the three moments which can be obtained by the weighted sum of each pixel value in the input unit circle [Tabatabai & Mitchell, 1984]. Lyvers proposed the subpixel edge detection based on spatial moments, which has a similar principle to the ones based on Gray moments, but a different moment operator [Lyvers et al., 1989]. This method uses six spatial moments to determine the parameters of edges from the 2D ideal step edge model, and locates the supixel edges by calculating the polar coordinates of edges. Another type of moment-based method is Zernike orthogonal moment-based method, which only use three moments instead of six as in spatial moments. In other words, the four parameters of the ideal step edge model are mapped into the three orthogonal moments with different orders and the subpixel edge position can be determined by these orthogonal moments [Ghosal & Mehrotra, 1993]. Although the moment-based methods are relatively complicated and have high computational cost, they also have outstanding properties. They are not sensitive to noise and quite accurate. Thus, they are applied in many significant aspects of image measurement [Yuanyuan *et al.*, 2010].

Curve-fitting methods utilize the edge points derived from conventional edge detectors at pixel level to the fit real edge shape curves with continuous functions, and then determine the subpixel edge positions. Usually, the least square method is adopted to estimate the continuous functions in transitional areas according to prior knowledge of an object such as shapes, sizes and positions in a scene. Certainly, a variety of fitting methods arise according to the different models. For example, fitting a line into spatial data points provided by the Canny operator, using a cubic spline or B-spline to get nonlinear borders of objects, and determining the parameters of the edge shape functions [Breder *et al.*, 2009][Yao & Ju, 2009][Bouchara *et al.*, 2007][Yueying *et al.*, 2007]. Clearly the position precision of curve-fitting methods strongly depends on the accuracy of edge points derived from the traditional edge detectors at pixel level. These methods are also sensitive to the points selected in transitional areas. Thus, good and accurate data points are need to guarantee exact subpixel edge positions.

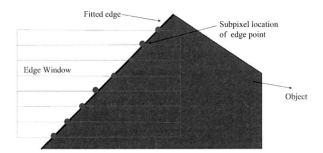

Figure 3.3: Subpixel edge based on curve fitting

Considering the measurement precision and noise, I employ a combination of reconstructive and curve-fitting methods in my work. An example of line edges is shown in Figure 3.3 where a border of the object is represented by the gray area. The red points indicate the detected edge points in transitional areas. Firstly, the 1D characters of intensity in transitional areas are reconstructed to determine the subpixel points at each scanning line. Then, a line is best fitted to the subpixel location in each scan line of pixels in the defined edge window. Finally, the whole line edge of the object is determined using the best-fit edge generated above. A more detailed analysis and description of subpixel edge processing is given in later chapters.

31

3.2 Stereo Vision Structure of Multilens Camera (eCley)

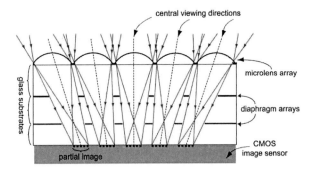

Figure 3.4: The working principle of eCley

The electronic cluster eye (eCley) adopts a compound eye solution to achieve the desired properties of the compact size, a large FOV and good sensitivity in the visible spectral range. Only a part of the whole FOV is captured in each channel and a final image is obtained by stitching all partial (channel) images by means of software processing. Figure 3.4 describes the working principle of an electronic cluster eye (eCley). I don't want to discuss the complicate mechanism in detail. It can be referred as in [Brueckner *et al.*, 2010]. I focus on how to form a digital image from incident rays and how to miniaturize the camera lens. There are five sequential channels displayed in Figure 3.4. Obviously, each channel only records a part of the whole scene as a partial image and a specific deviation angle is retained between the central viewing directions of adjacent channels. More accurately, definitely setting up the pitch difference between the optical component and the partial image of each individual channel. In order to obtain the specific angular sampling, the certain focal length (f_{sa}) is defined as $f_{sa} = p_{px}/\tan(\Delta\phi)$ where $\Delta\phi$ is the sample angle and p_{px} indicates the pixel pitch. Further the focal length of a single optical channel f can be derived from $f = f_{sa}/k$, i.e. the braiding factor k shortens the effective focal length while the imaging results still keeps closely with the single lens camera. Thus, this achieves the miniaturization of camera lens. Note that the interval between adjacent channels is necessary to maintain the acceptable light sensitivity and prevent optical crosstalk.

3.2.1 Stereo Properties of Adjacent Channels

The eCley has 17×13 channels with the same diameter, which are rigidly aligned with horizontal and vertical scanning lines of the sensor, namely the centers of each row or column always stay in the same line [Brueckner *et al.*, 2010]. Actually, the micro optics fabrication only produces some tiny tolerance, only in the order of magnitude of sub-micron. Even if a rather big deviation occurs by fabrication, they can be removed by the calibration process. Thus, two adjacent channels naturally comprise a stereo camera. So, is this stereo camera able to be used for measuring the distances of objects? Certainly, the answer is positive, but the method is quite different from the conventional one.

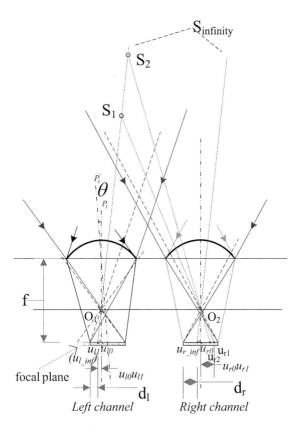

Figure 3.5: The principle of microlens stereo channels

Figure 3.5 shows the principle of imaging in two adjacent channels of the eCley based on the micro fabrication technique. Obviously, it is quite different from the canonical stereo camera that has two same cameras with the same size, focal length, and properties, especially parallel optical axes. But the structure of two adjacent channels changes the relative places of the image planes, i.e. the offset that results in the short focal length of the eCley.

Assuming both channels have the same focal length f and a different FOV, on the image plane, the point u_{l1} (here equal to u_{l_inf}) represents the image point of the object $S_{infinity}$ at infinity and the point u_{l0} denotes the intersection of the principal axis P_l and the image plane in the left channel. Naturally, their distance $u_{l0}u_{l1}$ is equal to the corresponding one in the right channel, i.e. $u_{l0}u_{l1} = u_{r0}u_{r\_inf}$. Let d_l indicate the distance between the left boundary of the left image plance and the point u_{l1} in the left channel. Meanwhile, let d_r indicate the analogous distance in the right channel. Observe $d_l - d_r \neq 0$, since there is the same size of image plane but different position of the center. Thus,

$$\Delta d = d_r - d_l = \tan \theta \cdot f \qquad (3.1)$$

where Δd represents the offset between two channels. If $\theta = 4°$, $f = 778\mu m$, then $\Delta d = 17 pixels$. That is the offset of two channels for the eCley.

3.2.2 Variation of Intensities Following Disparities in Boundaries of Objects

Detection and recognition of objects and their relative positions are the significant task to do in current real-time stereo systems that run on standard computer hardware. The locations of object borders (i.e. depth discontinuities) are quite important to retrieve proper object shapes for segmentation and recognition purposes [Hirschmüller et al., 2002].

From the previous analysis we know, when the distance between the camera and the object is over 86 mm, an accurate disparity is quite difficult to measure. It is usually seen as infinity. Actually, the disparity variation in one pixel is gradual (i.e. continuous transition), which reflects the object distance varies from the maximum valid distance (for example, 86 mm) to infinity. However, this transitional feature is connected with the intensity of the object's edge, and this fine feature can be interpreted by a lot of experiments and analysis of real images.

In Figure 3.6(a), a wing tip of the UAV (Unmanned Aerial Vehicle) called 'Psyche' was selected with a red rectangle in the gray image, which showed a distinguishable transition between the black wind tip and the gray sky. Figure 3.6(b) shows the close-up of the wind tip selected and it's gray values. There were about three pixels in the transition area, and it's gray values varied transitionally from dozens to hundreds. Essentially, the intensity variation reflects

(a) The image of UAV Psyche

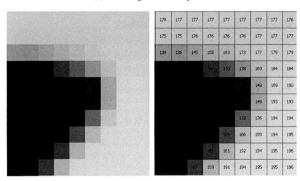

(b) The close-up and pixels value of Psyche's wing tip

Figure 3.6: UAV Psyche image and the close-up of it's wing tip

the position changes of subpixels, these relationship also was verified by the numerous tests.

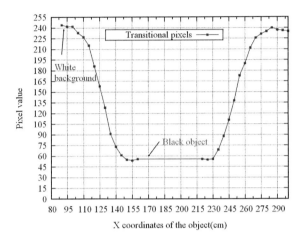

Figure 3.7: The relationship between the intensity and positions of subpixels

As seen in Figure 3.7, there were two transitional areas of the intensity on the both edges of the black square with the white background. Clearly the position of subpixels changed following the intensity. Although the whole transitional area did not vary linearly, the medial part of the transitional area could be linearly approximated, which provided a significant property for the stereo calibration and measurement approach.

If both the position of the pixel in the left image and the place of corresponding pixel in the right image are known for the canonical stereo camera, the disparity can be derived readily [Hartley & Zisserman, 2004]. Similarly, the subpixel disparity also can be obtained in the same way. From the prior study, we known the important property of the transitional area that the positions of subpixels changes follow the intensity. So, the intensity difference of a pair of subpixels must reflect the difference of their positions, namely the disparity of the subpixel pair. Further, the disparities of subpixels can be determined by the intensity difference in the transitional area of the object edge. Exactly this property causes the variation of the disparity of pixel pairs from the maximum to infinity.

3.3   Overview of Implemented Methods

In this thesis, I propose a method of obtaining depth information according to the relationship of intensity change and positions of key point points in transitional areas, which abstracts the subpixel positions of key points based on the edge properties. Now I give the overall frame of the proposed method for stereo vision in the eCley as follows.

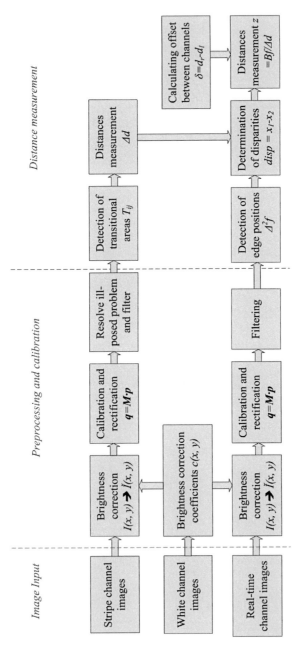

Figure 3.8: Subpixel distance processing diagram

As shown in Figure 3.8, there are two main routes: calibration of subpixel pixel pairs and distance measurement of objects running through the whole method. In the processes of calibration of subpixel pixel pairs, the idea is that the corresponding pixel pairs have fixed subpixel distances and they can be calibrated by an image sequence of the black-white triple model, which are captured by moving the tripe model along the horizontal direction. First of all, each pair of channel images shot at the stripe model should be corrected in brightness to eliminate the radiated deviation stemming from the different position in the channel. Then, the optical distortion and geometrical transformation are processed, followed by resolving the ill-posed problem and filtering. At this point the ideal channel images can be used to determine the subpixel distances of pixel pairs. The different approaches based on various features are employed to calibrate the distances in the transitional areas after detecting them. On the route of determining the distance measurement of the objects, the real-time channel images captured for various scenes are also corrected in brightness by the correction coefficients that are derived from the processing of the white board. Analogously, these necessary processes of calibration, rectification and denoise are executed before detecting the positions of the key points in transitional areas. Considering the real applications in various environments, an efficient and flexible algorithm of detecting edges is employed to attain the subpixel positions of key points. Certainly, the subpixel disparities of key points on objects can further be determined according to the calibrated distances of pixel pairs. Since the exclusive offset structure in eCley, the real baseline for each stereo corresponding pair also has a small difference at the subpixel level. Eventually, the depth information of objects can be obtained by the triangular relation among baseline, focal length and disparity.

39

Chapter 4

Correction and Calibration

4.1 Brightness Correction

According to the previous demonstration, disparity can be determined from the intensity difference of the transition area at the edge of objects. How to obtain this relationship? Assuming the variation satisfies some kind of function relationship: $D(.)$, the following equation is presented:

$$d = D(I_1 - I_2), \qquad (4.1)$$

where the disparity is described as d. I_1 and I_2 represent the intensities of the pixel 1 and the pixel 2 in the adjacent channels respectively. In a real environment, there are many factors to seriously affect the functionality such as the variation of illumination, differences of the scenes and deviations of channels. In this case, for obtaining the accurate difference of the intensity between a pair of sub-pixels, denoising and correcting intensity are very necessary before measuring the distances of objects with the eCley.

4.1.1 Intensity Correction Using Maximum Value

As the noise is inevitably caused by many factors in the real world, noise usually needs to be removed before processing the images. Certainly, many methods can be used, but in the stage of the sub-pixel calibration, the statistic mean is considered as a simply and feasible manner to get rid of the noise since it can also preserve the original intensity after the processing.

$$\bar{I}(x, y) = \frac{1}{n} \sum_{i=1}^{n} I_i(x, y) \qquad (4.2)$$

According to (4.2), the intensity mean of pixels, $\bar{I}(x, y)$, is derived from n image frames of the same scene and with the same parameters.

One important effect stemming from the miniaturization of the eCley, is that the brightness of each channel is not uniform, i.e. the parts near the rim of the channel look darker than the center compared to a uniform white board. Consequently, this difference of brightness evidently increases the uncertainty and measurement error. Especially the pixels that are far from the center of the channel must be corrected. The normalization without more complex processes is effective to reach the goal of correcting brightness.

$$c(x, y) = \frac{I_{max}}{I(x, y)} \tag{4.3}$$

Usually a pure white board is used in the correction. There must exist a pixel with the maximal intensity I_{max} around the center of the channel. Then, every gain coefficient of pixels, $\hat{I}(x, y)$, can be computed through the ratio of their original intensity to the maximal intensity.

4.1.2 Intensity Correction Using Voted Probability

A more sophisticated method is to employ the maximal probability of the intensity in the channel to estimate each correction coefficient. Assuming the point q_0 with maximal probability, which can be observed in the image, corresponds to the real white(or gray) point p_0, then the equation can be represented below by the normalization.

$$c(x, y) \cdot \frac{q_0}{\sum_{k=1}^{n} q_i} = \frac{p_0}{n} \tag{4.4}$$

where $c(x, y)$ represents the converted coefficient at the point (x, y), q_i is the ith pixel value, and n represents the number of pixels in the channel. Alternatively, (4.4) can be written as

$$c(x, y) = \frac{p_0 \cdot \sum_{k=1}^{n} q_i}{n \cdot q_0} \tag{4.5}$$

Once the corrected, or more accurately compensated, coefficients are determined, the brightness of the channel image can be corrected, and then each pixel has an even brightness, whatever its place in the channel is.

Figure 4.1 shows the brightness distribution in the channel before and after brightness calibration. In Figure 4.1(a), the whole distribution of brightness appears as an ellipsoid, i.e. there is the biggest intensity value near the center of the channel while the smallest values are located around the boundary of the channel. This non-uniform distribution of intensity would lead to the severe deviations in the distance measurement based on intensity. The objective of

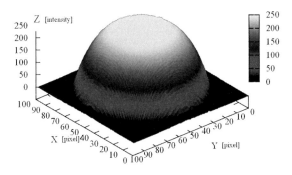

(a) Original channel brightness

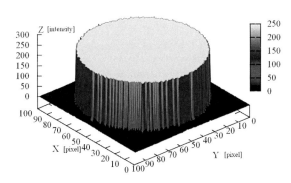

(b) Corrected channel brightness

Figure 4.1: Brightness distribution in the channel before and after brightness correction

brightness correction is to make the responding value of each pixel constant anywhere in the channel. For example, a result of brightness correction using voted probability is shown in Figure 4.1(b). Obviously the whole channel appears as a cylinder, which has an uniform brightness distribution even those pixels near the channel boundary keep the same intensity with others.

4.2 Offsets Between Adjacent Channels

The biggest difference of eCley and a common stereo camera is an offset with a constant angle for each channel. This property allows the eCley to exhibit super resolution in the context of a micro system structure. Usually, the offsets are measured at integer pixel precision. Actually they are different at the subpixel range. The difference of them affects significantly the measurement of subpixel disparities. Thus, the offsets between neighboring channels should be determined before measuring the subpixel disparities. How to attain these offsets? We can use the distance of the corresponding features at infinite distance to subtract the distance of centers of adjacent channels.

4.2.1 Channel Center Determination by Circle Fitting

When fabricating the eCley, the channel positions are devised accurately. This keeps all the channels in horizontal and in vertical alinement. But there are some biases for the final position of channels since the production process contains some errors and noise. Therefore, I utilize the devised position, i.e. the centers of channels, to be an initial value and then fit the accurate positions from the support points. In fact, the figures of channels are circles with the same radius. These points around the boundaries of channels can be employed to fit circles to the channels.

$$\varepsilon^2 = \sum_{i=1}^{n}(\sqrt{(x_i - x_0)^2 + (y_i - y_0)^2} - r) \qquad (4.6)$$

Equation (4.6) is the mathematic expression of determining subpixel pixel parameters. The estimated circle has the center (x_0, y_0) and the radius r. Accordingly, with the least square method, the optimal fitting will be reached at the minimal error. The implemented algorithm is described in the sequel. Firstly, selecting three non-collinear points around the boundary of the channel arbitrarily and using these points to fit circle $C(x_0, y_0)$ and it's radius r. Secondly, setting a threshold T_n in order to obtain a set of boundary points where all the points have smaller distances from the fitted circle and less than T_n. Afterwards, repeating the previous steps and then obtaining different fitted circles and the corresponding sets of boundary points. Finally, after sampling in time N images randomly, the optimal center and radius of the desired circle can be derived as the common circle for all the points around the boundary.

For obtaining the clear boundaries of channels, the channel image should be segmented from the camera image (whole image) as accurate as possible. Due to the white board versus the black back interval between channels, I can use a simple threshold to segment channel images. Certainly, the result depends on the threshold that can be derived from Otsu's method [Otsu, 1979]. Otsu's method is used to automatically determine the optimal threshold according to the shaped-based histogram of the image. In the case of white objects and black background, it can seek the best threshold, which gives rise to the minimal intra-class variance, to segment channel images and channel intervals to a binary image. As shown in Figure 4.2, the channel images

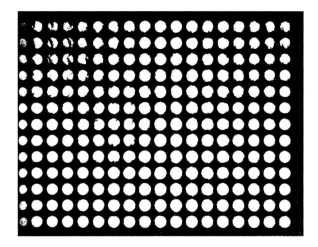

Figure 4.2: Binary channel images from the segmentation

are approximated to circles in the binary image after segmentation. The top left channel images are not complete circles due to shade and illumination effects in the original. The left column is not created by the channel images but by the reference holes for fabrication. Certainly, these special cases need to be observed when fitting circles.

The fitted results of top left channels are shown in Figure 4.3 where the white crossings near the centers of circles indicate the fitted positions of circles. In the channel $(0, 1)$, there are four points fitted as the center. No doubt all the three points except the point near center are far from the real center point, and I must eliminate these false points. The green circles are based on the fitted center points, which include the reference holes on the first column. The cyan circles represent the sets of center parameters without the reference holes. Nevertheless, there are some severe deviation points since the channel images are non-circles such as that in

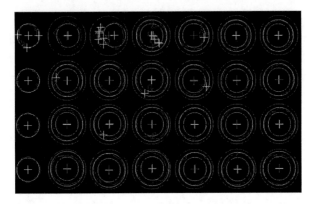

Figure 4.3: Circles fitted and their centers in the part of channels

channel (0, 2). The magenta circles show the valid fitted parameters and the positions of circles are also marked as magenta crossings at the centers of channels.

4.2.2 Centers Determination by Straight Lines Fitting

After fitting the circles, the coordinates of channel centers are really subpixel values. But they still are not accurate due to the deviation of the segmentation of the channel images. Thus, further processing is needed. We know that each channel is arranged in a straight line when the eCley was made. So, a line can be fitted to improve the accuracy of center coordinates. The linear fitting is implemented by

$$\hat{y} = ax + b \tag{4.7}$$

where a and b are the unknown parameters, the observed value x and y can be gotten from the circle fitting processing. Equation (4.7) is also called linear regression model. Given a data set $\{y_i, x_{i1}, \cdots, x_{ip} | i = 1, \cdots, n\}$ of n observed units, the regression values can be obtained by

$$\begin{cases} \hat{y}_1 = ax_1 + b \\ \hat{y}_2 = ax_2 + b \\ \quad \cdots \\ \hat{y}_n = ax_n + b \end{cases} \tag{4.8}$$

I define an error variable ε_i for the deviation from the observed value y_i to the corresponding regression value $\hat{y}$, and it is described as

$$\varepsilon_i = y_i - \hat{y}_i = y_i - (ax_i + b) \quad (i = 1, 2, \cdots, n). \tag{4.9}$$

46

Thus, the quadratic sum of all the deviations between observed values and regression line is set as S, which is derived from the following as

$$S = \sum_{i=1}^{n} (y_i - \hat{y}_i)^2 = \sum_{i=1}^{n} [y_i - (ax_i + b)]^2 \quad (i = 1, 2, \cdots, n). \tag{4.10}$$

The goal of regression is to make the value of S to reach a minimum regarding all the observed input values. Certainly I fulfill the task by the least squares method [Rao *et al.*, 2008]. Often the n equation from (4.9) are written in vector form as

$$\boldsymbol{y} = \boldsymbol{X}\boldsymbol{\alpha} + \boldsymbol{\varepsilon}. \tag{4.11}$$

where

$$\boldsymbol{y} = \begin{pmatrix} y_1 \\ y_2 \\ \vdots \\ y_n \end{pmatrix}, \boldsymbol{X} = \begin{pmatrix} x_{11} & x_{12} \\ x_{21} & x_{22} \\ \cdots & \cdots \\ x_{n1} & x_{n2} \end{pmatrix}, \boldsymbol{\alpha} = \begin{pmatrix} a \\ b \end{pmatrix}, \boldsymbol{\varepsilon} = \begin{pmatrix} \varepsilon_1 \\ \varepsilon_2 \\ \vdots \\ \varepsilon_n \end{pmatrix}.$$

Actually the second column of the matrix $\boldsymbol{X}$ is a constant vector whose elements all are 1. Using the lease square method, the solution of (4.11) is determined as

$$\hat{\boldsymbol{\alpha}} = (\boldsymbol{X}^T \boldsymbol{X})^{-1} \boldsymbol{X}^T \boldsymbol{y} \tag{4.12}$$

where $\hat{\boldsymbol{\alpha}}$ indicates the estimated parameter vector. The real discrete expression are given by

$$\begin{cases} a = \dfrac{\sum_{i=1}^{n} x_i \sum_{i=1}^{n} y_i - (2n+1)\sum_{i=1}^{n} x_i y_i}{\sum_{i=1}^{n} x_i^2 - (2n+1)(\sum_{i=1}^{n} x_i^2)} \\ b = \dfrac{\sum_{i=1}^{n} x_i \sum_{i=1}^{n} x_i y_i - \sum_{i=1}^{n} y_i \sum_{i=1}^{n} x_i^2}{(2n+1)\sum_{i=1}^{n} x_i^2 - (\sum_{i=1}^{n} x_i)^2} \end{cases} \tag{4.13}$$

From (4.13), the best fitting line can be obtained, which passes through the subpixel centers of channels. Note that more observed points lead to more accurate results that are affected less by noise. The whole camera includes 17×13 channels. All the initial coordinates of channels should be used. But the uncertainty of previous processing, i.e. the scattering points with big deviation, impacts the fitted results significantly. Ransac (Random sample consensus) is a rather robust algorithm for linear fitting [Hartley & Zisserman, 2004], which is a non-deterministic algorithm that estimates parameters of mathematic models by iteration from a observed data set containing outliers. The Ransac linear fitting algorithm is described as follows.

- select two points arbitrary and use them to derive a line l_1;

Table 4.1: Positions of centers and deviations

Column	X (pixel)	Y (pixel)	X-fitted (pixel)	Deviation (pixel)	Interval (pixel)	Ransac Deviation (pixel)
0	155.00	95.50	154.79	1.30	111.24	0.73
1	268.50	97.00	266.03	1.36	110.73	0.42
2	378.50	99.00	376.75	1.25	110.63	0.36
3	488.89	99.79	487.38	1.06	110.60	0.27
4	599.50	98.50	597.98	1.02	110.65	0.26
5	710.00	99.00	708.63	0.92	110.43	0.39
6	820.50	99.00	819.05	0.78	110.68	0.38
7	930.80	100.36	929.73	0.83	110.56	0.26
8	1041.00	101.00	1040.29	0.70	110.37	0.44
9	1151.71	100.41	1150.66	0.48	110.43	0.35
10	1261.51	100.59	1261.09	0.19	110.22	0.34
11	1371.88	101.35	1371.30	0.36	110.49	0.41
12	1481.90	101.40	1481.79	0.22	110.38	0.30
13	1592.00	101.50	1592.17	0.19	110.30	0.53
14	1702.81	101.67	1702.47	0.63	110.46	0.38
15	1812.50	102.50	1812.93	0.53	110.39	0.36
16	1923.00	101.00	1923.32	0.43	-	-

- compare the distance between all the points and the fitted line to a preset threshold T_h, then establish a new point set E_1, which contains all the center points of channels with a distance less than T_h;

- repeat the selection of two points randomly to get a different line l_i and its corresponding center points set E_i ;

- through N times sampling randomly, the line $\hat{l}$ is picked up as the ultimate fitted line for all the center points, which has the most points in the point set E_i ;

After deriving the horizontal and vertical lines, the crossing points of them must be the subpixel centers of channel.

The results of linear fitting for the first row of channels are listed in Table 4.1. Evidently, each coordinate is at a subpixel level. The deviations are quite different for the channels. They are affected by the binary segmentation of channel images. Here, I obtain the similar results to the ones in Figure 4.3, i.e. the first several center coordinates have quite big deviation since their channel images are roughly compared to others. However, these x coordinates become more accurate after using the Ransac method to get rid of the outliers, which are shown in the last column. Clearly, their deviations also decrease, i.e. all the positions of center points are optimal subpixel positions. The last two columns describe the subpixel intervals between neighboring channels, which means all the intervals are rather even except the first one.

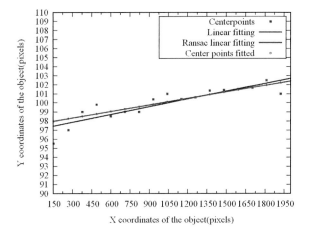

Figure 4.4: Fitting centers of channels

For comparing to the difference of linear fitted methods, I put the initial points, linear fitted and Ransac methods, and fitted points regarding the first row of channels into Figure 4.4 together. Obviously, the initial points scatter around the brown and blue lines, and several points on the left side are deviating a little bit far. These are the outliers of linear fitting. Therefore, the blue line that results from the Ransac method is more approximate to the real line than the brown line resulting from the common lease squares method. On the other hand, the root mean square of residuals is 0.87 for the least squares method while the smaller value 0.59 is for Ransac method. Additionally, the fitted center points are uniformly distributed on the blue line.

Now the intervals (distances) between two adjacent channels have been derived. Further the offsets are also obtained after determining the position difference of the two images from the same object at infinity. In nature, the objects that are farther than 8 m to the eCley can be considered as the infinite ones, which will be discussed in detail in Chapter 5.

4.3   Calibration and Rectification

Most applications in 3D computer vision need a prerequisite. I.e. it is necessary to obtain metric information from 2D images. Generally, the intrinsic parameters of the camera and the coefficients of distortions have to be measured. These processes are referred to as calibration in single cameras. Further processes such as the coplane computation of two imaging planes are referred to as rectification in stereo vision. A sophisticated method of calibration is from Zhang [Zhang, 2000], which robustly results in camera intrinsic parameters. Here I will only focus on calibrating distortions and implementing rectification.

4.3.1 Distortion of the Single Channel

In theory, the digital image formation model such as Lambertian and the camera model such as the pinhole model are assuming that the sensors perceive the illumination of scenes uniformly, and the optical components do not distorted the projection. In practice, however, the optical components, i.e. lenses, are not perfect due to manufacturing. On the one hand, most lenses are spherical in view of cost and convenience, which causes, essentially, the photometric distortion. On the other hand, it is very difficult to align the lens and the imaging sensor exactly. Therefore, calibration of lens distortion is a prerequisite of vision measurement. The two main photometric distortions are radial distortions arising from the fabrication of the lens and tangential distortions arising from the assembly process of the camera as a whole. Radial distortion is also known as barrel distortion, which causes rays farther from the center of a simple lens are bent too much compared to rays passing closer to the center. Whereas tangential distortion result from the case that the lens is not perfectly parallel to the image plane of camera. Both distortions can be analyzed by a mathematic model. Let (u, v) be the ideal pixel image coordinates without distortions, and (u_d, v_d) are the corresponding real observed image coordinates. According to the pinhole model, the ideal pixel points are the projection of the model of the real 3D points. In the camera coordinate system, similarly (x, y) and (x_d, y_d) are the ideal point's location (undistorted) and real location (distorted) respectively (which are normalized). The relation between them is described by

$$x_d = x + x[k_1(x^2 + y^2) + k_2(x^2 + y^2)^2] \tag{4.14}$$

$$y_d = y + y[k_1(x^2 + y^2) + k_2(x^2 + y^2)^2], \tag{4.15}$$

where k_1 and k_2 represent the coefficients of the radial distortion. The transformation from image space to camera space is given by

$$\begin{pmatrix} x \\ y \end{pmatrix} = \begin{pmatrix} f_x X^W/Z^w + u_0 \\ f_y Y^W/Z^w + v_0 \end{pmatrix}, \tag{4.16}$$

where the coordinates of object points in 3D space are defined as (X^W, Y^W, Z^w). I introduce two different lengths f_x *and* f_y that are actually the product of the physical focal length of the lens and the size of pixels($f_x = f \cdot dx$ *and* $f_y = f \cdot dy$), which are modeled as a rectangle with horizontal (dx) and vertical (dy) sizes. According to the pinhole model, I obtain $x_d = X^W/Z^w$, $y_d = Y^W/Z^w$.

Typically, the radial distortion is small in the common camera compared to fish eye cameras. So, some applications of vision such as object tracking only consider determining the intrinsic parameters of cameras by simply ignoring distortion [Zhang, 2000]. I can also use a similar strategy to obtain the distortion coefficients, i.e. to estimate k_1 and k_2 after having derived the ideal pixel positions by estimating the intrinsic parameters. By combining (4.15) and (4.16), the equation set for estimating the distortion coefficients is given by

$$\begin{pmatrix} (u - u_0)(x^2 + y^2) & (u - u_0)(x^2 + y^2)^2 \\ (v - v_0)(x^2 + y^2) & (v - v_0)(x^2 + y^2)^2 \end{pmatrix} \begin{pmatrix} k_1 \\ k_1 \end{pmatrix} = \begin{pmatrix} u_d - u \\ v_d - v \end{pmatrix} \tag{4.17}$$

Tangential distortion is the second-largest common distortion, which is caused by the deviations in the manufacturing process, i.e. the lens is not fully parallel to the imaging plane. The model of tangential distortion can be expressed as

$$x_d = x + [2p_1 y + p_2((x^2 + y^2)^2 + 2x^2)] \tag{4.18}$$

$$y_d = y + [p_1((x^2 + y^2) + 2y^2) + 2p_2 x], \tag{4.19}$$

where p_1 and p_2 are the coefficients of tangential distortion. Setting (4.19) into (4.17), I obtain the complete parameter equation set as follows:

$$\begin{pmatrix} (u - u_0)(x^2 + y^2) & (u - u_0)(x^2 + y^2)^2 & 2xy & 3x^2 + y^2 \\ (v - v_0)(x^2 + y^2) & (v - v_0)(x^2 + y^2)^2 & x^2 + 3y^2 & 2xy \end{pmatrix} \begin{pmatrix} k_1 \\ k_2 \\ p_1 \\ p_2 \end{pmatrix} = \begin{pmatrix} u_d - u \\ v_d - v \end{pmatrix} \tag{4.20}$$

In computer vision, we usually use vectors and matrix to simply express linear equations. Given m points in n images, all the equations from real observed points can be stacked together to obtain in total of 2mn equations, and its matrix form is expresses as $A\alpha = b$, where $\alpha = [k_1, k_2, p_1, p_2]^T$. Thus, I can use the methods of solving linear equations to estimate the desired parameters. In practice, the least squares method is usually used, and the linear least-squares solution is given by

$$\alpha = (A^T A)^{-1} A^T b. \tag{4.21}$$

4.3.2 Rectification of Two Stereo Channels

Almost all the stereo matching algorithms are based on the classical assumption that the images are captured by the canonical stereo setup, i.e. the real stereo setups have to be rectified. This means for stereo matching that the search space becomes only one-dimensional and that it is aligned with each scanning line. In practice, stereo systems do not really satisfy the canonical or standard stereo setup since there are individual deviations of the cameras and assembly errors. Thus, rectification is a necessary step in computation of stereo systems. Although the optical properties and physical parameters of channels are quite similar for different channels in an eCley due to the high precision fabrication, they might be thought as the same at pixel level. No doubt, a rectified system can further improve precision and reduce the errors.

Stereo image rectification is an essential process of image transformation in such as way that all the epipolars retain parallel in images and become collinear with each other and with the image scanning lines.In other words, the real stereo systems becomes the standard (rectified) stereo systems through the transformation. An inherent property of rectified systems is that the corresponding epipoles[1] in images move to infinity by transformation. Therefore, I can perform rectification by changing the positions of the epipolars in the images to infinity. In general, there are two steps to implement rectification: rotation of coplane and collineation.

- Rotating the left and right camera imaging planes and making them coplanar, which moves epipoles to infinity by a rotation matrix $\mathbf{Q}$, i.e. the epipolar lines become parallel.

- Rotating the epipolar lines to the direction of scanning lines and making them collinear, i.e. epipolar lines become parallel with scanning lines.

Without loss of generality, I use the stereo vision system with the same camera intrinsic parameters to describe rectification. As for the specific structure of eCley, I still employ the rectification from that common stereo system and only consider the offsets between two channels. In practice, the input data usually are two sets of corresponding image points in the left and the right image respectively. Certainly, I can derive the cameras intrinsic parameters and extrinsic parameters using the specific object point from Zhang's calibration method [Zhang, 2000], which provides the well known prerequisites for rectification by using the Bouguet's algorithm [Bouguet, 2010]. As shown in Figure 4.5, there is a real stereo system with the origin O_l and O_r. p_l and p_r are the image points of the object point P in the left and the right image planes respectively. First of all, I need to know the geometrical relationship between the two cameras. More accurately, how do I compute rotation R and translation T between the left and the right

---

[1]Since the centers of projection of the cameras are distinct, each center of projection projects onto a distinct point into the other camera's image plane. These two image points are denoted by e_l and e_r and are called epipoles. Refer to Figure 4.5.

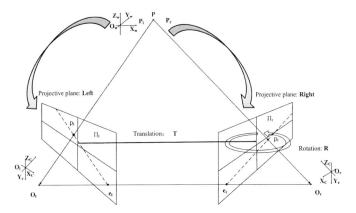

Figure 4.5: Calibration of two stereo channels

camera? I separately rotate and translate the object point P to get it's image points in the left and the right camera coordinates and obtain $P_l = R_l P + T_l$ and $P_r = R_r P + T_r$ for the two cameras respectively. Obviously, the transformation from the right image point P_r to the left image point P_l directly is described by $P_l = R^T(P_r - T)$, where R and T are, respectively, the rotation matrix and translation vector between the cameras. The relationship of the object point and its image points can be given by

$$
\begin{cases}
P_l = R_l P + T_l \\
P_r = R_r P + T_r \\
P_r = R(P_l - T).
\end{cases}
\tag{4.22}
$$

Solving the equations, I can derive the rotation vector and translation vector between left and right cameras as follows:

$$
\begin{cases}
R = R_r (R_l)^T \\
T = T_r - RT_l.
\end{cases}
\tag{4.23}
$$

Now, I discuss how to make two image planes coplanar. Since the problem is under-constrained, the choice is arbitrary. In general, I can create a transformation matrix $\mathbf{Q}$ by three mutually orthogonal unit vectors q_1, q_2 and q_3. Considering the left principal point O_l as the original, the vector q_1 is collinear with the translation vector $\mathbf{T}$ between the focus points of two cameras and described as [Trucco & Verri, 1998]

$$q_1 = \frac{T}{\|T\|}.$$

(4.24)

The second vector q_2 must be under-constrained, that is orthogonal to q_1. Clearly the direction to the principal axis along the image plane meets this requirement. The vector q_2 can be obtained using the cross product of q_1 by

$$q_2 = \frac{[-T_y\ T_x\ 0]^T}{\sqrt{T_x^2 + T_y^2}}$$

(4.25)

The third unit vector q_3 should be orthogonal to q_1 and q_2 simultaneously. It can be gotten using the cross product:

$$q_3 = q_1 \times q_2.$$

(4.26)

The orthogonal matrix that rotates the left camera around the center of projection in order to change the epipolar lines parallel and move the epipole to infinity is defined as

$$Q = \begin{pmatrix} q_1^T \\ q_2^T \\ q_3^T \end{pmatrix}.$$

(4.27)

The implementation of collineation is achieved by setting the following matrix:

$$R'_l = Q$$

(4.28)

$$R'_r = RQ.$$

(4.29)

R'_l and R'_r are new rotation vectors, which rotate left and right camera to reach a scanning line alignment. I can set each image point in the left camera as $P_l = [x, y, f]^T$; so the rectified image points in left camera coordinates are $R'_l P_l = [x', y', z']^T$. The corresponding rectified image points in left image coordinates are achieved as $P'_l = \frac{f}{z}[x', y', z']^T$ [Trucco & Verri, 1998].

In practice, I utilize the minimization of error between the image points and their reprojection points to obtain the optimal rectification parameters. This minimizes the resulting reprojection distortions while it maximizes the common viewing area. Bouguet provided the implementation of the algorithm in Matlab on the web [Bouguet, 2010], and the implementation in C++ version also can be found in the OpenCV library [Bradski & Kaehler, 2008]. I use a chessboard as the calibration object, of which I take images. For getting accurate results, the

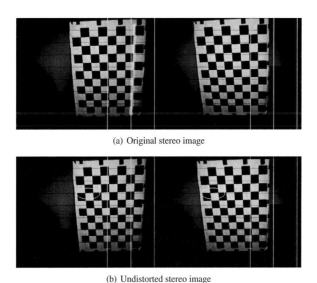

(a) Original stereo image

(b) Undistorted stereo image

Figure 4.6: Results from the calibration and rectification of a stereo camera

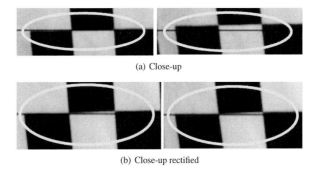

(a) Close-up

(b) Close-up rectified

Figure 4.7: Close-up before and after rectification

whole chessboard should be captured in two camera simultaneously and occupy an area as big as possible. I used a conventional stereo system to verify the algorithm, and then rectified the eCley with this method. The results are given as follows.

Figure 4.6 shows the images of the chessboard and the results from rectification. In Figure 4.6(a), the edges of black and white squares form many curves from bottom to top shown by the yellow line in the left image, and the scanning lines between left and right images are not

55

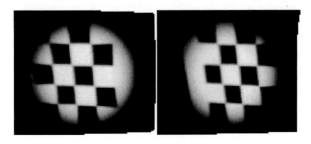

Figure 4.8: Rectified channel images of eCley

collinear. The rectified images are displayed in Figure 4.6(b). Here all the curves become straight and both rows in two images lie on the same scanning line.

The close-ups of a specific area from Figure 4.6 is shown in Figure 4.7. Obviously, we can see the rectified images are not only coplanar but also collinear. The scanning lines are placed correctly. The algorithm of stereo rectification reaches the goal of moving the epipoles to infinity. Thus, it can also be used in the eCley to implement stereo rectification. Notice that the corners detection should be done manually due to the low resolution. As described previously, the size of a channel image is only about 79 pixels × 79 pixels except for the offset between two channels in a channel image, which extremely reduces quality of the detected feature points and increases the difficulty of detecting corners precisely.

Chapter 5

Determining Subpixel Baselines of Channel Pairs

Based on Yuille's demonstration [Yuille, 1986], the edge points include all the information of the original image under the mean of different scales in the transitional areas. Therefore, we can recognize a 3D object according to these edge points which have most of the information of the image. On the other hand, the researches on biological vision systems have proven that vision neurons, which sense the vision as concentric circles on the retina, are able to detect the edges. In this chapter, we explore how to deal with the ill-pose problem on edges and the methods of detecting transitional areas, and the calibration of the subpixel baselines.

5.1   Resolving the Ill-Posed Problem

Usually, the points represent the respective physical meaning in different transitional areas. For instance, the points of transitional areas, also called edge area, may be the discontinuous points of surfaces. The transitional areas are formed by the overlap of surfaces or planes. Edge areas may arise at the overlap of two parts of different objects with different material or color; Transitional areas also occur at the overlap of an object and background. They may further be created at the borders of shades. Although there are different reasons for the formation of edges, a common property of edges is that a discontinuity of intensity occurs, i.e. an area with rapid change of intensity. At this position, the signal has many spatial high-frequency components at the edge. Therefore, all the detectors of transitional areas essentially detect high-frequency components. However, in natural images, signal and noise coexist in transitional areas. Thus they are difficult to detect. In other words, it's a fuzzy problem. Actually, edge detection is an ill-posed problem. The solution varies sharply when adding the high-frequency noise even with the low intensity. In the field of computer vision, regularization is usually applied to solve ill-posed problems [Tikhonov & Arsenin, 1977]. We define a observed signal as Y and the ideal signal as Z, so the relation of them can be given by

$$A(Z) = Y, \tag{5.1}$$

where $A(\cdot)$ is the transformed operator. However, Y may include noise and systematic errors due to the non-ideal observer. In the case of detecting features of a transitional area, Z may be the ideal intensity $I(x, y)$ and Y represents the real intensity. Only considering (5.1), the determination of Y from Z is obviously an ill-posed problem (generally A is irreversible)

$$E = \|A(Z) - Y\| + \lambda\|\rho(Z)\| \tag{5.2}$$

In fact, Y does not equal to $A(Z)$ arbitrarily due to errors, but it is the only evidence to determine Z. Thus, we expect to search the minimum of $\|A(Z) - Y\|$ with respect to Z. Meanwhile, Z represents the ideal physical signal and changes relatively slow compared to noise which can be described by a continuous function. On the whole, $\rho(Z)$ describes the smooth item with respect to the whole range of Z, which becomes smoother with smaller $\rho(Z)$. So the stable functional of $\|\rho(Z)\|$ should be small as well. λ is a constant factor that decides the weights of both items on the right of (5.2), and the research shows that bigger λ lead to a greater deviation of σ of filters.

In the application of detecting edges, the previous research [Chen & Hong Yang, 1995] proofs that the optimal solution of (5.2) can be attained by using a cubic B-spline function to filter the observed signal, i.e. the optimal smoothing filter is the cubic B-spline function. Due to the extreme similarity of cubic B-spline and Gaussian functions, we usually adopt the Gaussian function as the smoothing filter to get an unbiased estimation of the ideal edge signal. According to the above analysis, the method of regularization further verifies that a filter is a prerequisite for the detection of edges and the cubic B-spline or the approximate Gaussian function are the optimal filter.

A definite mathematical model is necessary to describe edge positions. As stated previously, the points of edges indicate positions with drastic changes in transitional areas approximately. For example, a one-dimensional signal is shown in Figure 5.1. The magenta step signal indicates the ideal edge model, but the real edge signal is similar to the brown curve with a gradual change because the physical signal does not jump from one status to another. The red and blue curves represent the first order differential and second order differential of the real edge signal. Clearly, the position of 0, which is defined as edge position, is the common feature point of all the curves. The maximum of first order differential and the zero crossing point of second order differential. In 2D images $I(x, y)$, its first order derivative should be thought as the modulus gradient. The

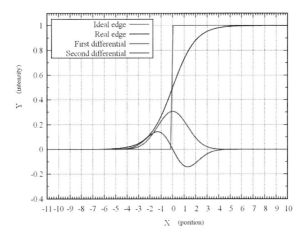

Figure 5.1: Signal edge and its differential

gradient vector is given by

$$\boldsymbol{g} = \nabla I(x, y) = \left(\frac{\partial I(x, y)}{\partial x}, \frac{\partial I(x, y)}{\partial y} \right)^T. \tag{5.3}$$

On the other hand, the second derivative of edges is considered as the directional derivatives along the gradient directions. For reducing the complexity, we usually employ the Laplace operator, a second order differential operator with isotropy, which is expressed by

$$\nabla^2 I(x, y) = \frac{\partial^2 I(x, y)}{\partial x^2} + \frac{\partial^2 I(x, y)}{\partial y^2}. \tag{5.4}$$

The isotropy results from that $\nabla^2 f$ is an invariant under rotation. When detecting edges, the derived second derivative from the Laplace operator is independent to the directions of image edges [Pratt, 2001]. Summarily, determining the local maximum of the first derivative and the zero crossing point of the second derivative are basic approaches for edge detection. Although both approaches are equivalent in the outcome, the method of detecting the zero crossing point is quite simple and easy to implement. As we know, noise gives rise to the ill-posed problem for edge detection. So, the smoothing filter is devised to get rid of noise before extracting edges. Assuming the pulse transfer function of a filter as $s(x)$, first of all, we use $s(x)$ to filter noise in the signal and get the filtered signal $g(x) = I(x) \otimes s(x)$; $\otimes$ denotes the convolution operation. In this way, we derive the edge points from first or second derivative of $g(x)$ [Cyganek & Siebert,

2009]. Since the order between the convolution operation and filter can be exchanged due to commutativity as follows:

$$g'(x) = \frac{dI(x) \otimes s(x)}{dx} = \frac{d}{dx} \int_{-\infty}^{+\infty} I(t)s(x-t)dt$$

$$= \int_{-\infty}^{+\infty} I(t)s'(x-t)dt = I(x) \otimes s'(x),$$

(5.5)

we can combine the two processes of filtering and computing the derivative, and then obtain the first differential filter $s'(x)$ and the second differential filter $s''(x)$. Thus, the general method of edge detection is to devise a smooth filter $s(x)$, and detect the local maximum of $I(x) \otimes s'(x)$ or the zero crossing point of $I(x) \otimes s''(x)$.

$$g_\sigma(x) = \frac{1}{\sqrt{2\pi}\sigma} e^{-\frac{x^2}{2\sigma^2}}$$

(5.6)

$$g'_\sigma(x) = \frac{-x}{\sqrt{2\pi}\sigma^3} e^{-\frac{x^2}{2\sigma^2}}$$

(5.7)

$$g''_\sigma(x) = \frac{1}{\sqrt{2\pi}\sigma^3} e^{-\frac{x^2}{2\sigma^2}} \left(\frac{x^2}{\sigma^2} - 1\right)$$

(5.8)

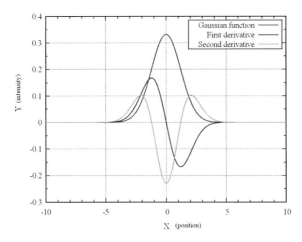

Figure 5.2: Gaussian function and its differential

The common smoothing filters are based on the Gaussian function or its derivative, which are described in (5.6), (5.7) and (5.8); where σ denotes the standard deviation of the Gaussian function. A bigger σ causes a stronger smoothing. But actually, a too big σ gives rise to eliminating the sharp high-frequency components of edges even if most noise is filtered. This is the biggest challenge in edge detection. In Figure 5.2, the red, blue and green curve indicate the Gaussian function of (5.6), first differential Gaussian function of (5.7) and second differential Gaussian of (5.8). Compared to Figure 5.1, the function curves based on Gaussian are similar to that of edges and its derivatives, i.e. we can use these model based on Gaussian function to approximate the real edges.

5.2 Detection of Transitional Areas

The significant property of transitional areas are edge points that imply rich features of objects and scenes. As the definition of edge exists at both intensity and gradient levels, we can detect edges or transitional areas at intensity or gradient of images.

5.2.1 Edge Detection by Intensity

According to the definition of edge, the position of an edge is placed at the points with maximal variation of intensity. How to find the maximal variation? According to Figure 3.6 and Figure 3.7, the intensity mainly varies at a few pixels in the transitional area, especially around the mean intensity between the object and the background. An outstanding property of the transitional areas is that the variation of the intensity always appears monotonously and the difference of the intensity between the adjacent pixels is typically bigger than that at other places in the image. How do we decide where the transitional area is? So, this property can be employed to find the transitional areas. A row of pixels at the transitional area from Figure 3.6 is shown

Figure 5.3: A piece of the transitional area

in Figure 5.3, which was taken from a black object with the white background. Obviously the pixels $p_3 - p_6$ lie in the transitional area, and their pixel values change sharply from the 54 to 193. This variation spans at least 25 gray levels between two adjacent pixels, whereas the pixels in the object or background vary only in several gray levels, what is usually caused by noise. Thus, (5.9) is introduced to represent the transitional area. Namely, consecutive pixel values are monotonous in a row of the image and their differences are greater than a preset threshold between arbitrary adjacent pixels.

$$T_{ij} : \begin{cases} p(x_i, y_0) < p(x_{i+1}, y_0) \quad or \quad p(x_i, y_0) > p(x_{i+1}, y_0) \\ |p(x_i, y_0) - p(x_{i+1}, y_0)| > \alpha \quad i = 1, 2, ..., n, \end{cases} \tag{5.9}$$

where $p(x_i, y_0)$ represents the value of the pixel x_i at the specific row y_0; $p(x_{i+1}, y_0)$ is the adjacent pixel; α is a preset value that distinguishes the domain pixels in the transitional area T_{ij} with the size of n.

5.2.2 Detection from Signal Gradient

The other main method of detecting edges is to utilize the gradient information. Let us assume that the function I(m, n) represents discrete values of intensity at the image point with coordinate(m, n), which corresponds to a 2D continuous function I(x, y). Both the discrete and the continuous forms of the luminance function can be converted to each other by sampling and interpolation. With these assumptions we can use the intensity gradient vector defined in (5.4) [Cyganek & Siebert, 2009]. Each image point can be described as a small patch of the triple (x, y, I(x, y)), where the normal vector is defined as [Klette *et al.*, 1998]

$$\mathbf{n} = \left[\frac{\partial I(x, y)}{\partial x}, \frac{\partial I(x, y)}{\partial x}, 1 \right]^T . \tag{5.10}$$

Thus, the edge detection modulus of the gradient and normal to the gradient vectors are derived from

$$\|\nabla \mathbf{I}\|_{L_2} = \sqrt{\left(\frac{\partial \mathbf{I}}{\partial x}\right)^2 + \left(\frac{\partial \mathbf{I}}{\partial y}\right)^2} \tag{5.11}$$

$$\|\mathbf{n}\|_{L_2} = \sqrt{\left(\frac{\partial \mathbf{I}}{\partial x}\right)^2 + \left(\frac{\partial \mathbf{I}}{\partial y}\right)^2 + 1}. \tag{5.12}$$

Generally, the L2 norm is applied to the modulus and the vector. For simplification of computations, the L1 norm commonly is adopted by

$$\|\nabla \mathbf{I}\|_{L_1} = \left|\left(\frac{\partial \mathbf{I}}{\partial x}\right)\right| + \left|\left(\frac{\partial \mathbf{I}}{\partial y}\right)\right|. \tag{5.13}$$

According to the definition of edges, the modulus of the gradient vector grows in transitional areas with much variation of the luminance signal. More accurately, it happens just in the case of edges as well. Therefore, we can compute the gradient by means of some methods of discrete differentiation such as the Sobel operator to extract edges precisely in the images. The Sobel operator is one of the discrete differentiation operators for image processing, which focuses on edge detection by computing an approximation of the image gradient. Using the Sobel at an arbitrary point of an image results in the corresponding gradient (or normal vector). According to the previous analysis of filters, the derivative image is represented by the convolution of the image and a differential filter. The Sobel operator uses two 3×3 derivative masks, one for horizontal and one for vertical derivative, It is convolved with the image $\boldsymbol{I}$ to approximate the derivatives. The image $\boldsymbol{G_x}$ with horizontal derivative approximation and $\boldsymbol{G_y}$ with vertical derivative approximation are given as

$$\boldsymbol{G_x} = \begin{pmatrix} -1 & 0 & 1 \\ -2 & 0 & 2 \\ -1 & 0 & 1 \end{pmatrix} \otimes \boldsymbol{I} \text{ and } \boldsymbol{G_y} = \begin{pmatrix} -1 & -2 & -1 \\ 0 & 0 & 0 \\ 1 & 2 & 1 \end{pmatrix} \otimes \boldsymbol{I}. \qquad (5.14)$$

The final gradient and gradient direction of the image can be determined by combining the resulting gradient approximations $\boldsymbol{G_x}$ and $\boldsymbol{G_y}$ using:

$$G = \sqrt{\boldsymbol{G_x}^2 + \boldsymbol{G_y}^2} \qquad (5.15)$$

$$\theta = \arctan(\frac{\boldsymbol{G_y}}{\boldsymbol{G_x}}), \qquad (5.16)$$

where θ indicates the angle of direction. When the edge is vertical, θ becomes 0 degree, and a horizontal edge corresponds to the value of 90 degree.

To implement the Sobel operator, we deduce an approach of discrete pixels. The differential of each image point with respect to the direction x and y is approximated as follows:

$$
\begin{aligned}
\frac{\partial I(x,y)}{\partial x} &\approx \frac{I(x+1,y) - I(x-1,y)}{2} & \frac{\partial I(x,y)}{\partial y} &\approx \frac{I(x,y+1) - I(x,y-1)}{2} \\
&\approx \frac{I(x+1,y-1) - I(x-1,y-1)}{2} & &\approx \frac{I(x-1,y+1) - I(x-1,y-1)}{2} \\
&\approx \frac{I(x+1,y+1) - I(x-1,y+1)}{2} & &\approx \frac{I(x+1,y+1) - I(x+1,y-1)}{2}
\end{aligned}
$$

$$(5.17)$$

From (5.13), we can obtain the L1 norm of the gradient. It is approximate to the sum of two direction differentials regarding x and y. Further, the equation (5.13) is replaced by (5.17).

$$
\begin{aligned}
\|\nabla I(x,y)\|_{L_2} &\approx \|\nabla I(x,y)\|_{L_1} \\
&\approx \frac{1}{4}\left(4\left\|\frac{\partial I(x,y)}{\partial x}\right\| + 4\left\|\frac{\partial I(x,y)}{\partial y}\right\|\right) \\
&\approx \frac{1}{4}\Big(2\big(I(x+1,y) - I(x-1,y)\big) + \big(I(x-1,y+1) - I(x-1,y-1)\big) \\
&\quad + \big(I(x+1,y+1) - I(x+1,y-1)\big)\Big)
\end{aligned}
$$

$$(5.18)$$

Obviously, the gradient of images can be attained robustly and accurately using the above (5.18), which utilizes the differences between positions of the key point and its neighborhood points and weight the contribution of each point in its neighborhood.

5.3 Calibration of Subpixel Baselines

In general, we only consider a uniform baseline for a classic stereo camera. The disparity is derived from the difference of corresponding pixel positions between the left and the right image. But this uniform baseline is not correct for channel pairs in the eCley, i.e. each channel pair has its own baseline that is usually not equal to others at subpixel precision level.

5.3.1 Subpixel Referred Position Based on Local Mean

In the sequel we call pixel position that are handled with subpixel precision simply subpixels. According (4.1), the difference between the two corresponding subpixels in both adjacent channels can be determined easily if the function $D(.)$ is known, but, the feasible approach actually is to transform (4.1) to (5.19), that is

$$
\begin{cases}
\bar{d} = \frac{1}{n}\sum_{k=1}^{n}(x_{1k} - x_{2k}) \\
I(x_{1k}) = I(x_{2k}) \qquad I(x_{1k}), I(x_{2k}) \in M_{ij}
\end{cases}
$$

$$(5.19)$$

where the disparity $\bar{d}$ is represented by the difference of positions between a pair of subpixels (x_{1k}, x_{2k}) that are with the same intensity. Here a question arises, namely $I(x_1) = I(x_2)$. That is: which points should be picked out for determining the disparity? Simply the neighborhood of the mean intensity in the transitional area, namely the range M_{ij} in (5.19), is quite suitable for computing the disparity since the medial area is more sensitive to a variation of position. Certainly, the effect factors will emerge when all the subpixels referred are adopted, especially those subpixels far from the mean. A feasible method is using the local subpixels to calculate the mean instead of all the subpixels in the transitional area. The results are thereby called as local mean that is more accurate than the mean derived from all the subpixels.

5.3.2 Subpixel Edge Fitting

Actually, the intensity distribution curves of transitional areas are quite similar to the Sigmoid function [Zhang et al., 2009b]. The sigmoid function also called the S function is widely applied to neural networks, due to its continuity, smoothness and monotony. The Sigmoid model of an intensity distribution in transitional areas is created by

$$I(x) = \frac{a}{1 + e^{\frac{-(x-b)}{c}}} + d, \qquad (5.20)$$

where a denotes the difference between maximal and minimal intensities, b and d indicate the offsets of the S function at x-axis and y-axis respectively, and the gradient of the edge is represented as c. For determining the subpixel positions, let be $y = I(x)$ and derive the first derivative of the intensity by

$$y' = I'(x) = \frac{(y-d)(a-y+d)}{ac}. \qquad (5.21)$$

When the derivative of (5.21) with respect to y is attained, let it be equal to zero as follows,

$$\frac{dy'}{dy} = \frac{a + 2d - 2y}{ac} = 0. \qquad (5.22)$$

Based on (5.22), as $y = d + \frac{a}{2}$, the responding point reaches the maximum, which is exactly the desired position. Here, the x-axis coordinate of the maximum should be equal to b.

According to the previous demonstration, the first differential feature of transitional area looks like the Gaussian curve. Thus, Gaussian function can be used to fit the gradient character of transitional area [Fabijaska, 2012]. The Gaussian model of an intensity gradient distribution is described as (5.7).

The proposed algorithm is summarized below. First of all, the brightness deviation of the input image is corrected by (4.5). Noise has been eliminated through the mean of multiple frames, Afterwards, every transitional area is detected. Then, to find the translation of the object pattern, fitting curves of the intensities distribution and obtaining the subpixel edges are implemented by (5.19) and by (5.20) respectively. Finally, the subpixel distances of stereo pixel pairs are determined using the difference between subpixel edge positions.

5.4 Experiments and Analysis

Through the previous statements, we have obtained the approaches to the detection of transitional areas and calibration of the subpixel baselines. The following experiments will verify their reliability, accuracy, and robustness in real environments.

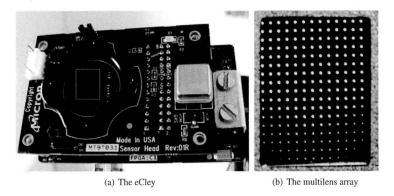

(a) The eCley   (b) The multilens array

Figure 5.4: Artificial compound eye: the eCley

In the experiments, we employed a demonstration system of eCley that is made of the Aptina MT9T031 USB high resolution camera and a micro lens array. The related technical data are listed as follows:

Table 5.1: Parameters of the eCley demonstration system

Property	Value
Image sensor:	Aptina MT9T031
Size of image:	2048 by 1536
Size of channels :	17 by 13
Focal length:	778 micron
Pixel pitch :	3.2 micron
Channel pitch:	355.2 micron (111pixels)
Visual field:	7.8 degree
Angular distance:	0.2 degree

The demo version of eCley is shown in Figure 5.4(a). The red rectangle area marks the micro lens array, whose close-up is shown in Figure 5.4(b). Clearly, the micro lens array with 17 by 13 units is closely glued to the sensor. This definitely decreases the focal length to be smaller than 1mm. Apart from the necessary eCley, an object with some clear transitional areas is needed to provide a target in the experiment. In our experiments, two classes of objects: synthetic objects and real objects, are used to verify the performance of subpixel algorithms and calibrate the subpixel distances of pixel pairs. Additionally, the main parameters of the computer are the CPU pentium 4 3.20 GHz and 1.50 GB of RAM.

5.4.1   Experiments Using Synthetic Objects

Generally, subpixel algorithms are needed to verify their precision and reliability. For reaching the goal, an ideal edge is required, which has known position and low noise. A simple and

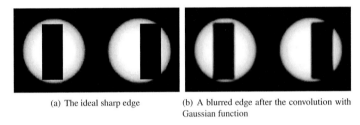

(a) The ideal sharp edge   (b) A blurred edge after the convolution with Gaussian function

Figure 5.5: The ideal sharp edge and a blurred edge

easy solution is to generate a straight line edge from a black object and white background. We adopted a pair of white channel images as background, which were shot at a white paper or white wall. Meanwhile, two black blocks were added in the two channel images as objects, which were put at a fixed positions in both channel images [Zhonghai *et al.*, 2002]. Figure 5.5(a) shows the synthetic channel images. The first edge lies in (1024.5, y) in the left image, where y indicates the row coordinate of the edge. Correspondingly, the column coordinate of the first edge is at 1024.5 in the right channel image. The fixed distance between the two edges is 270 pixels.

According to the formation of a digital image, we know that the object edges in digital images captured by a digital camera can be see as the results from the convolution of the real object edges. This processing can be described as:

$$I(x) = f(x) \otimes g(x), \qquad (5.23)$$

where the digital edge image $I(x)$ is derived from the convolution of the real edge signal $f(x)$ and the operator $g(x)$. Figure 5.5(b) shows the simulated edges that were generated by convoluting the ideal edges in Figure 5.5(a) with the Gaussian function. The advantage of simulated edges is to keep the same edge position as an ideal edge even if they have different transitional areas. Furthermore, the distance between the edges in the left channel image and its corresponding edges in the right channel image still keeps the same, i.e. 270 pixels.

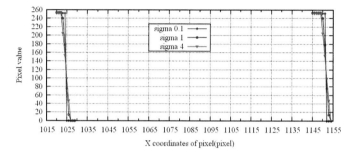

Figure 5.6: The ideal edge and its blurred edges with different σ

In fact, for the simulated edges derived from the Gaussian function, the transitional areas of edges depends on the standard deviation σ of the Gaussian function. As Figure 5.6 shows the red curve represents the ideal edge that appears sharp, the blue and magenta curves are the simulated edges after the Gaussian blur. The edges have the same position, i.e. a common intersection point at the transitional area. This shows the significant property that the range of the transitional area increases as σ rises up.

We used the simulated edged image (Figure 5.5(b)) to examine the proposed algorithms. Both approaches—based on local mean and based on subpixel edge fitting—are examined respectively. Part of results shows Table 5.2, where the ideal distance is 127 pixels from the vertical edge at 1014.5 pixels in the left channel to the one at 1151.5 in the right channel. The first column indicates the y coordinates of edge points from 756 to 775. The edge points estimated by the Sigmoid curve fitting are listed in the third and fifth columns. From the last two columns, we know the distances of corresponding edge points in both images are quite close to the ideal distance of 127 and their distance means are near equal. This means that the local mean method and fitting method are able to implement the distance detection of stereo pixel pairs and the results reach high subpixel precision (their standard deviations are around 0.01). Note that the estimated positions of edge points don't significantly affect the determined distances since the position errors in the left channel are compensated by the one in the right channel. For example, the position error of edge point at (1024, 765) in the third column is 0.054 pixel ($1151.5 - 1151.446 = 0.054$), which can be compensated by the position error (also 0.054 pixel) of the corresponding point (1151, 765). Thus, the distance between the stereo pixel pair is still the same 127 pixel like the ideal distance.

Table 5.2: The results of distance measurement using local mean and subpixel edge fitting

Y-pos. (pixel)	X-pos.-ch1 (pixel)	Position-ch1 (pixel)	X-pos.-ch2 (pixel)	Position-ch2 (pixel)	Distance-F (pixel)	Distance-LM (pixel)
756	1024.5	1024.454	1151.5	1151.446	126.993	126.993
757	1024.5	1024.451	1151.5	1151.446	126.995	126.995
758	1024.5	1024.447	1151.5	1151.446	126.999	127.005
759	1024.5	1024.446	1151.5	1151.446	127.001	127.001
760	1024.5	1024.445	1151.5	1151.446	127.001	127.000
761	1024.5	1024.445	1151.5	1151.446	127.001	127.000
762	1024.5	1024.446	1151.5	1151.446	127.000	127.000
763	1024.5	1024.446	1151.5	1151.446	127.000	127.000
764	1024.5	1024.446	1151.5	1151.446	127.000	127.000
765	1024.5	1024.446	1151.5	1151.446	127.000	127.009
766	1024.5	1024.446	1151.5	1151.446	127.000	127.013
767	1024.5	1024.446	1151.5	1151.446	127.001	127.013
768	1024.5	1024.445	1151.5	1151.446	127.001	127.013
769	1024.5	1024.454	1151.5	1151.446	126.993	126.993
770	1024.5	1024.455	1151.5	1151.446	126.991	126.993
771	1024.5	1024.459	1151.5	1151.446	126.988	126.983
772	1024.5	1024.459	1151.5	1151.446	126.988	126.983
773	1024.5	1024.460	1151.5	1151.446	126.986	126.982
774	1024.5	1024.462	1151.5	1151.446	126.985	126.979
775	1024.5	1024.464	1151.5	1151.446	126.983	126.976
Mean	1024.5	1024.451	1151.5	1151.446	126.995	126.996
Std.	-	0.006	-	-	0.006	0.011

The noise resisting ability is a very important for evaluating the performance of subpixel algorithms. By adding random noise in the simulated edge images, we can obtain the simulated image with noise. This examines the performance of methods for measuring subpixel distances. We attained results similar to Table 5.2. The comparison of the results with noise and the ones without noise are shown in Figure 5.7, where the red curve represents the deviations between the ideal distance and estimated distances according to the fitting method with respect to no noise environment. Correspondingly, the blue curve describes the results with respect to the noise environment. Obviously, the medial part of the blue curve, i.e. around the y coordinate of 763 pixels, is rather smooth and with small deviations, but at far edge points away from the center at pixel 763 such as pixel 774 relatively big deviations arise. This is due to the fact that the distant pixel points are distorted more severely than the ones near the optical center. Although the

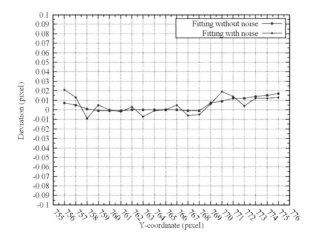

Figure 5.7: The effect of noise to the proposed algorithms

results from the environment with noise have small fluctuation around the red curve, the whole variation range of the blue curve is limited to 0.02 pixel. This lets the fitting method sustain high precision also under noise. In addition, the local mean method obtained a similar performance in our experiments. Summarily, the proposed methods to measure the subpixel distances of stereo pixel pairs have a strong noise resisting ability and high measurement precision.

5.4.2 Experiments Using Real Objects

Although the proposed algorithms perform quite well at the images with a synthetic object, we still have to test the subpixel distances of pixel pairs in a real environment.

For producing the accurate edges in real images, the object adopted was a flat board with black and white stripes, which produced several vertical transitional areas. Figure 5.8(a) shows the experimental setup. The object plane was kept parallel to the plane of the camera. Note that the distance between the eCley and the object should be over the minimum with respect to the minimal precision. In our experiments, we estimated the ideal precision smaller than 1% of a pixel, i.e. the minimal object distance should be kept over 8636mm (derived from (eq1.22) and the minimal distance 86mm with respect to the disparity of one pixel). Certainly, the eCley needs to stay in a fixed position and the stripe board should be moved at small steps on a straight line with a range of 1000mm, so that the edges of black and white could sweep over all the pixels in a specific channel. A captured image is shown in Figure 5.8(b) where we specified the stereo channel is the channel 6-8 (row-column) and 6-9. For improving the

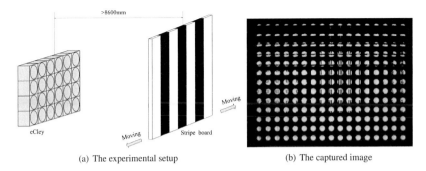

(a) The experimental setup   (b) The captured image

Figure 5.8: Experimental setup and captured image

efficiency of test, we designed three pairs of black-white stripes with six edges. Certainly, an image with more pairs of black-white stripes could be designed theoretically. But the diffraction of edges, the resolution of the image and the field of view must be considered.

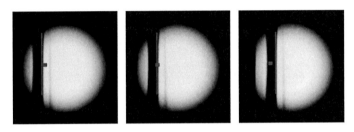

Figure 5.9: Obtaining the intensity of a specific pixel with respect to an edge

For obtaining the intensity properties of transitional areas, we need to record the intensity of the specific pixel when an edge passes at a specific pixel position following the movement of the stripe pattern. As shown in Figure 5.9, for example, the red pixel point indicates a specific pixel and the blue line describes an edge that passes by the specific pixel point. From the left image to the right image, each intensity of the specific pixel is stored. After the edge passes by the specific pixel, the whole information of the transitional area is attained, which can be used to compute the distance of stereo pixel pairs.

According to the previous analysis, the sub-pixel disparity, theoretically, can vary in the range of 100% to 1% of the size of a pixel, when the distance between the object and the eCley lies in the range 86mm to 8636 mm. Therefore, the test distance for experiments should be kept over the 8636 mm. Actually a distance around 8–9 m is enough because the image of the object is too small to be identified clearly with respect to the extremely far distance.

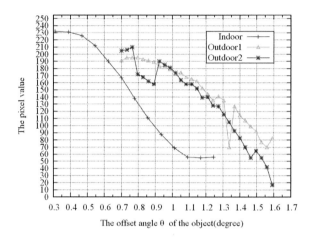

Figure 5.10: The property of transitional area tested outdoors vs. indoors

Additionally, an even illuminance is a very significant requirement, which will make the experiment a success or a failure. For instance, we used the setup as shown in Figure 5.8(a) and shifted the stripe pattern along the horizontal line that parallels with the image plane of the eCley. While the other parameters remained unchanged, we captured a sequence of images for each position of the stripe pattern indoors and outdoors. Figure 5.10 shows the outcomes from outdoor and indoor tests respectively. Obviously, the result from the indoor test was quite smooth, i.e. only little noise was measured. The curve of the transitional area rather resembles the theoretical one mentioned in Section 2. However, the two curves (blue and green) tested outdoors were extremely irregular and some aberrant points arose in the curves as well, which stemmed from the uneven daylight. Note that the intensity range of transitional area is changed by the different luminance in scenes. In Figure 5.10, the result from the indoor tests is in the range from 56 to 233 but the outdoor results vary in different ranges. Certainly, a large range of intensity for transitional area is necessary to get highly accurate results.

Generally, the effect of thermal noise can't be eliminated in digital cameras. The formed images have slight variations in intensity at different times. This slight variations can not be ignored in case of signal processing at subpixel level. A practicable method is to use multiple frames. The number of the frames, n was set at 10 frames. If we would set n too small, the noise in the images could not be removed effectively. If it is set too big, this would aggravate the computational cost. The accuracy of results was considerably affected by the intervals of the relative movement between the object and the eCley. Assuredly, small intervals increased the number of sampled points. This resulted in a smoother curve and less errors. In view of the size of the stripe pattern and the test distance, the interval of object movement distance was kept below 50 mm. Within this range, satisfying results could be obtained.

Due to the special structure of the eCley, the brightness of a channel image is not consistent with input brightness, that is, the brightness at the border is darker than the one in the middle of the channel. So, the brightness deviation must be corrected at first in the experiments. 10 frames of a uniform white object were captured by the eCley at proper position, and data nearly without noise were obtained. Further, corrected coefficients were computed thought (4.5), which could be used to correct the brightness of input images and let all the points have the same response to a homogenous white object. One of the uncorrected and corrected results is showed in Figure 5.11(a), which was shot at a homogenous white paper in a scene with uniform illuminance. Obviously, the corrected right image has a homogenous intensity at each pixel in the whole channel. However, the intensities of pixels in the uncorrected left image are not uniform, especially near the border. We extract a scan line from the channel images, which are marked by red and blue lines respectively in Figure 5.11(a). As shown in Figure 5.11(b), the channel image had a deviation of the brightness between the center and the border at the same scanning line as the red curve shows. The corrected blue curve becomes regular as a square wave. Both sides of it reach the same grey level as the middle. Analogously, the magenta curve indicates the corrected brightness using the same coefficients at the different illuminance and distance. To verify the performance of the brightness correction in a complex environment, two channel images that were shot under uncorrected brightness and corrected brightness respectively shown in Figure 5.11(c), evidently the corrected right image lights the edge of the channel comparing to the left one. This is, brightness correction not only corrects the brightness deviation from optical imaging, but also enlarges the visible range of the channel.

For utilizing the information of the neighbor rows of the image as well, the transitional band needs to be vertical during the movement of the stripe model. In this case, the information of the neighbor rows would not affect the intensity of the measured row but just eliminate the noise and improve the robustness. Note that the threshold α in (5.9) was set at 20, which depends on the test condition such as the illuminance, object distance, and size of the object. As part of results shown in Figure 5.12, there were several sets of data tested at different positions where the transitional areas of the stripe model were moved to. Definitely, the four transitional areas from the white to black area were picked out to show the properties of the channel. From Figure 5.12(a), the axis x indicate the diameter of the whole channel from pixel 1108 to 1191. Obviously, the red curve occurred near the boundary in the current channel, especially the blue curve also represented the boundary of the adjacent right channel that lies near the pixel 1302 though the transitional area arose in the middle of the current channel, and the rests of green

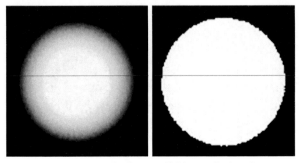

(a) The uncorrected and corrected brightness of a channel

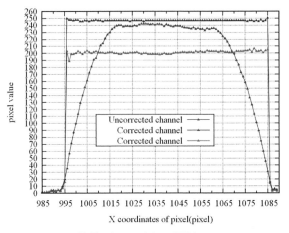

(b) The changes of channel brightness

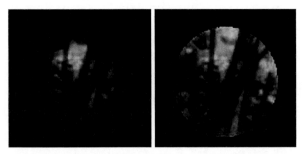

(c) The effect of brightness correction to real channel images

Figure 5.11: The comparison of the brightness of channels

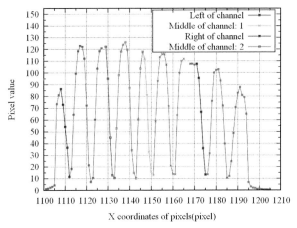

(a) Transitional characters at different positions

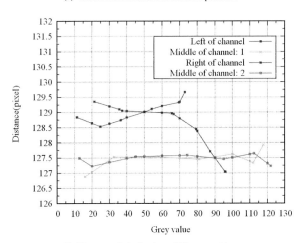

(b) Distances of pixel pairs at different positions

Figure 5.12: The comparison to calibrated stereo pixel distances at different places

and magenta curves were in the middle of the test channel. Correspondingly, the measured results in Figure 5.12(b) show that the pixels at the middle position of the channel had stable values, particularly near the mean of the transitional area. Although the boundary curves were inevitably affected by the distortion, the neighborhood of the mean, i.e. from 40 to 60, still stayed stable and had a constant mean.

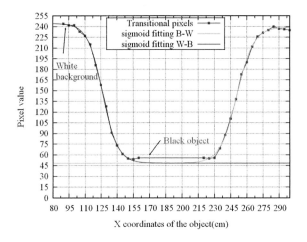

Figure 5.13: Fitting the sigmoid curves to the edges

On the other hand, we further used the fitting method to get subpixel distances. Commonly, this data fitting method adopts non-linear least square, which can easily deal with noise and derive accurate and robust results. In our experiment, the GNU Scientific Library (GSL) that is a numerical library for C and C++ programmers was used to implement the tasks of curve fitting. For the fitting questions, a key point is to set up the initial values. According to (5.20), we set a, b, c, and d to 250, 1250, 7, and 10. Then, the algorithm will convergence to the desired precision after a few steps. Note that the number of pixels out of the transitional area considerably affects the fitting results. Therefore, the accurate determination of the transitional area is very important for the fitting results. Figure 5.13 shows the results from two transitional areas, where the red line indicates the detected feature points in the transitional areas, and the blue and green curves represent the fitting curves from the sigmoid model. Obviously, the fitting curves employ the information of all the feature points very well regardless if it is a white-black edge or a black-white edge.

Consequently, the sub-pixel distances between the pixels in the current channel and the corresponding pixels in the adjacent channel were measured completely. Note that all the distances

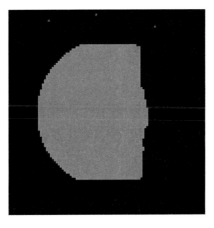

Figure 5.14: The valid region existing the stereo pixel pairs between two adjacent channels

of pixel pairs can be measured, but not all pixels have corresponding pixels in the adjacent channel at infinite object distance. An area with stereo pixel pairs was shown in Figure 5.14. Clearly, the top and bottom of a channel don't have stereo pixel pairs. As well, at a part of the right region of a channel don't have subpixel correspondences from the stereo pixel pairs. These just verified the previous analysis again, which indicated that only two third of the area of a channel was validly useable for stereo matching. This is due to the fact that both channels have different viewing angles. Table 5.3 lists the part of the sub-pixel distances of pixel pairs in a specific row (Y=765). Certainly each pixel pair had a different distance, and the pixel pairs at the middle of the channel such as X=1029, 1032 and 1035 presented the closest value 128.3, but a big difference arose near the boundary, especially at the pixel of (1054,765). On the whole, all the distances measured by the proposed method had small variances, that is, the measured distance is quite stable and precise.

For observing the subpixel distance distribution in a channel, I constructed a 3D image to demonstrate the character on the whole. As shown in Figure 5.15(a), X and Y indicate the related positions of the channel and Z denotes the number of pixels where 0 denotes 124 pixels, 1 denotes 125 pixels, and so on. Figure 5.15(b) shows the surface of subpixel distance variation in the channel where the subpixel distances can be read from Z axis. Whether in Figure 5.15(a) or in Figure 5.15(b), we can see that the subpixel distances are not a uniform value, i.e. each stereo pixel pair has its own subpixel distance(relative viewing angle). Commonly, big distances arise at the border of a channel. The central area of a channel has even distances compared to the border.

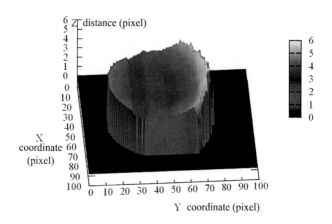

(a) Subpixel distances distribution in a channel

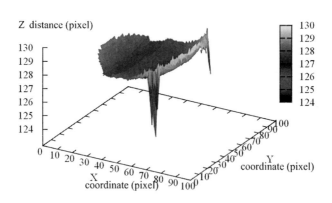

(b) The surface of subpixel distance distribution

Figure 5.15: The distribution of subpixel distances of stereo pixel pairs

Table 5.3: The measured distances of the channel and their deviations

Position X (pixel)	Y axis (pixel)	Mean (pixel)	Variance (%)
1014	765	127.375	2.4
1017	765	127.225	2.5
1020	765	128.912	4
1023	765	127.759	0.5
1026	765	127.682	1.2
1029	765	128.195	1.1
1032	765	128.461	1.2
1035	765	128.32	1.5
1038	765	128.1	1.4
1041	765	127.853	1.5
1047	765	127.706	1.1
1051	765	125.954	1.14
1054	765	126.363	1.52

Utilizing the calibration distances of stereo pixel pairs, the eCley can implement the measuring of distances of objects, and 3D position. Generally, the baseline in a stereo camera may be considered as uniform for all the pixels. Actually, I have shown that for each stereo pixel pair, their (angle) distance can be seen as the subpixel baseline in a corresponding pixel pair which is different from the others. This distance belonging to each stereo pixel pair provides a prerequisite to extend the stereo function in eCley. In the next chapter, I will discuss how to apply the calibrated subpixel distances of stereo pixel pairs in the stereo channel to measure the distances of objects.

Chapter 6

Real-Time Subpixel Distance Measurement for Objects

For finding the correspondence of two images, there are two categories of processing methods in real applications. One is based on intensity patterns, and the other is to use the features of images. In general, the edges of objects are more easily adopted by object recognition rather than points and surfaces. The main reason is that edges include many properties of objects such as contour and intensity difference. In this section, we will develop a real-time method to determine subpixel based distances of objects.

6.1 Overview of the Approach

The implemented procedure is described as follows. From the diagram: firstly, a pair of channel

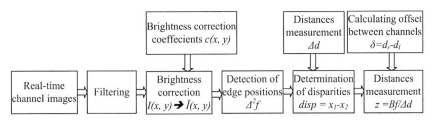

Figure 6.1: The diagram of implementation of measuring distance for objects

images is captured. It is preprocessed by a filter to eliminate noise and calibrate the optical distortion. Then the brightness correction coefficients obtained from the previous chapter are applied to balance the brightness in the whole channel image. Afterwards, the pure image is used to detect the edge positions of objects. The subpixel disparity between stereo pairs of

81

pixels is computed. It results from the difference of the determined subpixel baseline correction (see previous chapter) and the distance of pixel pairs in the image. Finally, combining the input offset based on adjacent channels, the subpixel disparity and the baseline of pixel pairs, the measurement of the subpixel distances of objects is reached successfully.

6.2 Image Preprocessing

An image from a real environment generally includes many factors such as noise and optical distortion. In many cases the scene is not invariable, thus the previous methods that calculates the mean from multi-frames of the same scene is not exactly appropriate any more. Actually, the noise in images can be described by the Gaussian distribution. As mentioned before, the ideal filter for the edges that include much high-frequency information is the Gaussian filter, i.e. the Gaussian filter is an optimal estimation. The one-dimensional Gaussian filter is described by

$$g(t, \sigma) = \frac{1}{\sqrt{2\pi}\sigma} e^{-\frac{t^2}{2\sigma^2}}, \tag{6.1}$$

where t denotes the argument that can be along with x axis or y axis. Analogously, the 2D Gaussian kernel $G(x, y, \sigma)$ is given by the formula:

$$G(x, y, \sigma) = \frac{1}{2\pi\sigma^2} e^{-\frac{x^2+y^2}{2\sigma^2}}. \tag{6.2}$$

It is observed that (6.2) can be expressed with the multiple between two one-dimension Gaussian kernels in the formula:

$$G(x, y, \sigma) = g(x, \sigma)g(y, \sigma), \tag{6.3}$$

where x, y denote two free coordinates and σ is a parameter. That means that the Gaussian kernels are isotropic, i.e. their characteristic is perfectly symmetric in all directions. We can commonly employ the 2D Gaussian function to get rid of noise directly, but this way causes high computational cost. In the context of a real-time application, two one-dimensional functions (6.2) are preferable over the 2D filter.

6.3 Detecting Coarse Edges of Objects

Since the shapes of real objects are not as simple as a line or a circle but more complicated, the transitional area cannot be directly determined by the method of the previous chapter. Here we employ another method to estimate rough edge positions of objects in a real-time environment.

6.3.1 Detection of boundaries of Objects

John F. Canny proposed the Canny edge detector that uses a multi-stage algorithm to detect a wide range of edges in images [Yao & Ju, 2009]. Due to the excellent performance of the Canny operator, this type of edge detection is popular at pixel level. The Canny operator is an optimal edge detection algorithm which tries that not only to mark as many real edges in the image as possible but also marks edges as close as possible to real position and tries to avoid false detections. On the whole, the stages of the Canny algorithm consist of noise reduction, finding the intensity gradient of the image, non-maximum suppression, and tracing edges through the image based on two thresholds.

Noise reduction: In order to eliminate the effect of noise, the Canny edge detector employs the Gaussian filter, i.e. the raw image is convolved with the Gaussian filter ((6.1) and (6.3)). Essentially, there is a tradeoff between filter noise and edge detection, but the result of Gaussian filtering keeps the edge properties better than other denoising approaches.

Determination of the intensity gradient: In view of the gradient of the image, we can approximate it using the first finite difference so that the two partial derivative matrices (G_x and G_y of the image at the respective direction x and y can be obtained. We devise a simple convolution operator as

$$s_x = \begin{pmatrix} -1 & 1 \\ -1 & 1 \end{pmatrix}, \; s_y = \begin{pmatrix} 1 & 1 \\ -1 & -1. \end{pmatrix} \qquad (6.4)$$

Thus, the partial derivative of each image point with respect to the gradient direction x and y is obtained as follows:

$$G_x(x,y) = (f(x,y+1) - f(x,y) + f(x+1,y+1) - f(x+1,y))/2 \qquad (6.5)$$
$$G_y(x,y) = (f(x,y) - f(x+1,y) + f(x,y+1) - f(x+1,y+1))/2, \qquad (6.6)$$

and then the edge gradient and direction can be derived from (5.16).

Non-maximum suppression: A fact is that the greater value in the gradient matrix response is the greater gradient value of the point in the image. But it doesn't mean that point is the desired edge point. Given estimates of the image gradients, a search is carried out to determine if the gradient magnitude assumes a local maximum in the gradient direction. In other words, that is to search a local maximum and set the values of points with non-maximum to zero so most non-edge points can be rejected. As we can see in Figure 6.2, for the goal of non-maximum suppression, we firstly need to judge whether the gradient value at the pixel C is the maximum in its neighborhood of eight pixels. The blue line indicates the gradient direction at C, the definite fact is that the local maximum of C must exist on the blue line. In other words,

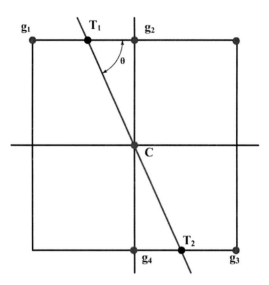

Figure 6.2: The principle of non-maximum suppression

besides the point C, the other two points (T_1 and T_2) where the gradient direction line interact with the adjacent scanning lines are also candidates for the local maximum. Therefore, we can know whether the point C is exactly the local maximum in its neighborhood by comparing C with T_1 and T_2. Assuming the gradient as data value of C is less than T_1 or T_2, that means the point C is not the local maximum and C can not be the edge point. Note that T_1 and T_2 have to be calculated by interpolation since only the values of pixels such as g_1, g_2, g_3 and g_4 are derived directly in the neighborhood. The non-maximum suppression results in a binary image in which the edge points have a constant value such as 127 and all others are set to 0. By now, the rough edges have been detected, but there is still some noise and false edges in the binary image. Thus, further processing is still needed for finding the final fine edges.

Tracing edges through the image based on two thresholds: Large intensity gradients are more likely to correspond to edges than small intensity gradients, but they also include many false edges. Therefore, we need to filter the false edges by thresholds. In most cases, a high threshold can suppress most false edges to a related pure edges image, i.e. edges with little false information, but the edges are likely to be unclosed due to the high threshold. Thus, a low threshold is usually used in the Canny algorithm. The fine edges are extracted by

$$
I(x,y) = \begin{cases} 1, & (I(x,y) > T_h) \text{ or } (T_h > I(x,y) > T_l \text{ and } I_k > T_h), \ k \in [m,n] \\ 0, & \text{others} \end{cases} \quad , \quad (6.7)
$$

where T_h and T_l denote the high and low thresholds respectively and $I(x,y)$ represents the gradient value of arbitrary pixel. In the neighborhood [m,n] of $I(x,y)$, the kth pixel's value is I_k.

6.3.2   Decision of Transitional Areas

Using the Canny edge detector, the edge positions of objects can be found. However, these edges only describe the coarse positions of real edges, i.e. at pixel resolution level. For the goal of subpixel positions, the decision of transitional areas becomes the significant procedure since this does not only define a range for the subpixel computation but also gets rid of outliers. As we

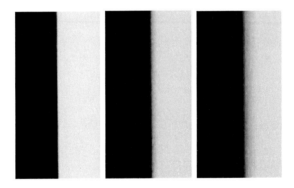

Figure 6.3: The edge of black and white under different focusing

know, the edge appears differently with respect to changing focusing, which causes variations of transitional areas. A real example can illustrate this property of the camera. We used a common camera with fixed focal length to take several photos under different focusing for a black straight object with white background. In order to obtain images with different values of blurring parameters, we shot some images with different manual focusing. To avoid artifacts the uncompressed images were converted from original raw format to 24 bits color PNG format. We only focus on the properties of edges, a cutout of 240 x 120 pixels around the edge (Figure 6.3) was extracted for analysis. Certainly, all the color channels are not necessary for out task.

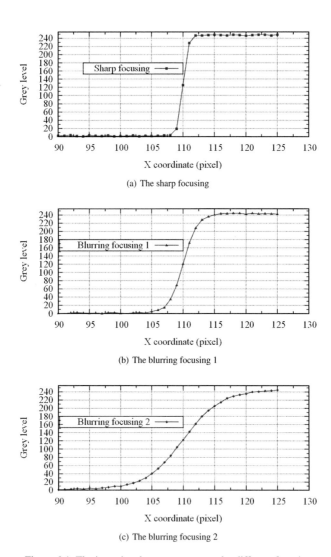

(a) The sharp focusing

(b) The blurring focusing 1

(c) The blurring focusing 2

Figure 6.4: The intensity character curves under different focusing

Here the information of red channel was adopted in our test.

Figure 6.4 shows the plot of the edges under different focusing. Obviously, a big blurring parameter generated a big transitional area, while a sharp edge makes its transitional area more narrow. Based on this property of camera, we can confirm that a camera with fixed focal length and fixed focusing distance imaging an arbitrary object will create edges of all objects to have a fixed size of the transitional area. Thus, this property provides great convenience for specifying the transition area. The definition of transitional area is described by

$$I(x, y) = \begin{cases} 1, & x \in [x_0 - m, x_0 + n] \\ 0, & \text{others,} \end{cases} \tag{6.8}$$

where $I(x, y_0)$ denotes the value of the pixel (x, y) and x_0 is the position of an edge pixel that was determined by the above Canny edge detector. m and n are the preset number of pixels for the above and below limit of a neighborhood at the position of coarse edges.

6.4 Distance Measurement at Subpixel Level

According to (4.1), the difference between the two corresponding subpixel positions in both adjacent channels can be determined easily if the function $D(.)$ is known, but, the feasible approach actually is to transform (4.1) to (6.9). That is, the disparity $\bar{d}$ is represented by the difference of subpixel positions between a pair of pixels (x_{1k}, x_{2k}) that have the same intensity, namely $I(x_1) = I(x_2)$.

$$\begin{cases} \bar{d} = \frac{1}{n} \sum_{k=1}^{n} (x_{1k} - x_{2k}) \\ I(x_{1k}) = I(x_{2k}) \qquad I(x_{1k}), I(x_{2k}) \in M_{ij} \end{cases} \tag{6.9}$$

Here a question arises, that is, which points should be picked up for determining the disparity? Simply the neighborhood of the medial intensity in the transitional area, namely the range M_{ij} in the (6.9), is quite suitable for computing the disparity. There are two benefits: one is that the middle of the edge area is not sensitive to the variation of places, the other one is that the intensity of pixels around the middle value approximately vary linearly. So that the linear interpolation can be used to obtain the required data. Therefore, based on the points around the middle of the left channel, the corresponding points with the same intensity can be obtained by linear interpolation in the right channel. For instance, when the difference between the object and the background is 100, the value 50 should be selected as the base point. The base point's neighborhood is exactly specified by extending one third of the value 50 as the limit of the neighborhood, i.e. range[33, 66]. Then, in this range, the mean of the intensity regarding n pairs of points is derived from the linear interpolation. Considering the effect of noise, the bilinear interpolation is an ideal approach to derive the corresponding points since it can not only eliminate the noise but also remove the deviation of the response of sensors. The (6.10) describes the real implemented method.

$$I(x,y) = \begin{pmatrix} x_2 - x & x - x_1 \end{pmatrix} \cdot \begin{pmatrix} I(x_1,y_1) & I(x_1,y_2) \\ I(x_2,y_1) & I(x_2,y_2) \end{pmatrix} \cdot \begin{pmatrix} y_2 - y \\ y - y_1 \end{pmatrix}, \qquad (6.10)$$

where the four points (x_1, y_1), (x_1, y_2), (x_2, y_1), (x_2, y_2) are contiguous, and the intensity $I(x, y)$ at the point (x, y) can be computed by their corresponding positions.

The advantage of bilinear interpolation is that the four points around the unknown point can be utilized to support the value of the expected point rather than only depend on the gray value of a point. Thus, the bilinear interpolation actually appears better that the nearest neighbor interpolation. However, it will result in edges with big error when the edges have big curve rates or big incline with respect to the vertical scanning lines.

Certainly, a bicubic interpolation would not only use the information of adjacent vertical points but also the rate of change between the adjacent points. Thus, the precision of edge positions could be improved very much. But the complexity and its high computational cost restrain its application in real-time environment.

An alternative approach is the cubic polynomial interpolation that considers the rate of change between the adjacent points and computes efficiently. The cubic polynomial interpo-lation utilizes the four points on the same scanning line and that they are sequential in position. We can obtain definitely the interpolation points by

$$L_4(x) = \sum_{i=0}^{4} \prod_{j=0}^{4} \frac{x - x_j}{x_i - x_j} f(x_i). \qquad (6.11)$$

In the above equation, the left side $L_4(x)$ indicates the Lagrange interpolation, and the right side is the sum of four items that are the product of base functions of interpolation and function values with respect to each point. The cubic polynomial interpolation is highly precise and efficient in real applications.

6.5   Experiments and Performance Analysis

In general, the subpixel algorithms need to be tested and verified in ideal and real environment. In the experiments, we first used the synthetic object to examine the algorithm for measuring real-time subpixel distances. Then, the real performance of the proposed method was verified in indoor and outdoor environments in order to implement the desired results.

6.5.1   Verifying Effectiveness and Precision of Algorithms

As in the previous chapter, the algorithm for measuring real-time subpixel distances should be examined using a synthetic object. We adopt the simulated edge image of Figure 5.5(b), which has the edge positions of 1024.5 pixels and 1152.5 pixels in the respective channel. The ideal distance between the two vertical edges is 127 pixels. The experimental results are listed in Table 6.1.

Table 6.1: The results of distance measurement and deviations

y-pos. (pixel)	x-pos.-ch1 (pixel)	x-pos.-ch2 (pixel)	Distance (pixel)	Distance-noise (pixel)	Deviation (pixel)	Deviation-noise (pixel)
756	1024.5	1151.5	126.995	126.989	0.005	0.011
757	1024.5	1151.5	126.997	126.985	0.003	0.015
758	1024.5	1151.5	127.005	126.992	-0.004	0.008
759	1024.5	1151.5	127.002	127.007	-0.002	-0.007
760	1024.5	1151.5	127.001	127.004	-0.001	-0.004
761	1024.5	1151.5	127.000	127.003	0.000	-0.003
762	1024.5	1151.5	127.000	126.986	0.000	0.014
763	1024.5	1151.5	127.000	126.992	0.000	0.008
764	1024.5	1151.5	127.001	127.003	-0.001	-0.003
765	1024.5	1151.5	127.006	126.990	-0.006	0.010
766	1024.5	1151.5	127.006	127.003	-0.006	-0.003
767	1024.5	1151.5	127.011	127.005	-0.011	-0.005
768	1024.5	1151.5	127.013	126.990	-0.013	0.010
769	1024.5	1151.5	126.993	126.991	0.007	0.009
770	1024.5	1151.5	126.995	126.981	0.005	0.019
771	1024.5	1151.5	126.984	126.990	0.016	0.010
772	1024.5	1151.5	126.983	126.988	0.017	0.012
773	1024.5	1151.5	126.983	126.976	0.017	0.024
774	1024.5	1151.5	126.982	126.976	0.018	0.024
775	1024.5	1151.5	126.988	126.978	0.012	0.022
Mean	1024.5	1151.5	126.997	126.992	0.003	0.008
STD	-	-	0.009	0.010	-	-

Clearly, we find that the distances between the feature points and their corresponding points are quite stable. More accurately, the distances vary around the ideal distance of 127. The standard deviations of distances are rather small (about 0.01); even in the case of the simulated edge image where noise was added. The last two columns show the deviations from the ideal distance, which demonstrates the small values. This means that the algorithm for measuring real-time subpixel distances can effectively fulfill the subpixel distance measurement and obtain highly precise outcomes. Certainly, the goal of the chapter is not only to derive distances of

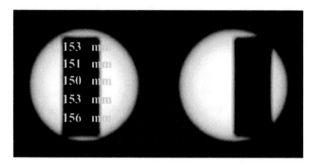

Figure 6.5: The estimated distances of the synthetic object

stereo pixel pairs but also to determine object distances. Here, we utilize the calibrated distances of stereo pixel pairs to get the disparities of objects' edge features. Further, the object distances from the eCley can be determined by (2.3). Figure 6.5 shows the distances between the feature points of the synthetic black object and the eCley. Although determining the real distance of the synthetic object is meaningless, the results demonstrate that the subpixel differences between each pixel pair in the stereo channels are different instead of the uniform value as the baseline of stereo channel. The distances of near pixels are quite close so that the eventual object distances that are estimated from the feature pixel points of the object are approximate.

6.5.2 Tests Indoors and Outdoors

In our experiments, we used the eCley to measure the distances of preset objects indoors and outdoors, and then evaluated the precision of the proposed algorithm comparing to the ground truth. The influence factors of the measurement precision are also discussed at the end, such as the effect of offset error, brightness correction, distortion, noise et al.

In the indoor experiments, we used three objects: a tea box, glue and a stripe board as the scene. They had preset distances to the eCley. Theoretically, the eCley can test the distance with respect to subpixel disparity from the minimum distance corresponding to one pixel up to infinite. Usually, the error is around $1/20$ pixel for a camera with the resolution of 512×512 [Cyganek & Siebert, 2009], i.e. the distance measurement at the range over $1/20$ pixel

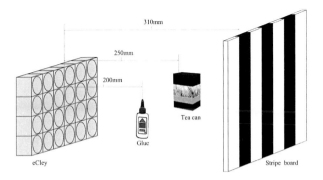

Figure 6.6: The ground truth of distances

disparity is of no concern. But for far distances with respect to the very small disparity big errors occur. This is caused by the difference between the real camera and the pin-hole model with the illumination, noise and blooming effect. Therefore, the preset distances have been in the measurable range of [90 mm, 900 mm] in our experiments. The objects should have rather obvious edges distinct from the background to be detected using the proposed algorithm. We adopted a tea box with sharp edges and a glue bottle with smoothly curved edges. In addition, the stripe board used was an ideal object for comparison with others because it has highly contrasted black and white stripes. For avoiding the effect that different distances results from different positions of the same object, we keep the eCley vertical with all the objects. The set of experimental objects and the eCley are shown in Figure 6.6.

The coarse edges are obtained by the Canny detector that has been discussed in the previous section but the result of Canny detector includes strong edges and weak edges. Therefore, we need to filter the weak edges to improve the accuracy of the distance measurement. Additionally, the horizontal edges and inclining edges also need to be eliminated. Since the processing for the whole image including all the channel images, the rough edges include the rims of the channels. Certainly, the rims of the channels should not arise in the purified channel images.

Figure 6.7 shows an edge image of the channel at row 6 and column 8 in the eCley. By the Canny detector, the channel image of Figure 6.7(a) becomes the edge image of Figure 6.7(b), in which there are some horizontal edges, heavily inclining edges and the rim of the channel. Figure 6.7(c) and Figure 6.7(d) are the results without the rim and horizontal and inclining edges.

Table 6.2 lists the ground truth and the results from the several measured points on the edges of the objects. Clearly, each sampled point on the edge leads to a different real distance around the ground truth, which is caused by many factors such as illumination, sensor sensitivity and

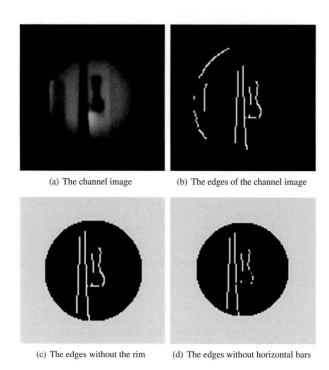

(a) The channel image   (b) The edges of the channel image

(c) The edges without the rim   (d) The edges without horizontal bars

Figure 6.7: Determination of edges and transitional areas

Table 6.2: The measured distances of the channel and their deviations

Objects	Ground truth (mm)	Point1 (mm)	Point2 (mm)	Point3 (mm)	Point4 (mm)	Point5 (mm)	Mean (mm)	Deviation (%)
Glue	200	211	208	204	202	199	205.4	2.5
Tea box	250	262	260	256	256	251	251	3.7
Stripe board	310	324	321	318	314	312	316	4.2

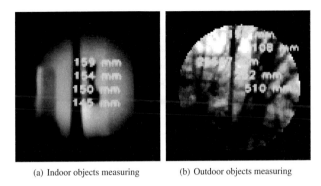

(a) Indoor objects measuring   (b) Outdoor objects measuring

Figure 6.8: The results of measuring object distances indoors and outdoors

noise. In fact we can see is that the mean values of the distances are quit near to the ground truth. The error is smaller than 5%. For single measurement, it is also smaller than 5%. Even if there are different types of edges, the error of the measurement is limited to 10% of our experimental range, i.e. the proposed method fulfills the goals of the subpixel measurement. Additionally, the precision of measurement decreases for distant objects but stays in the mentioned range.

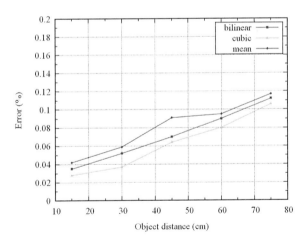

Figure 6.9: Comparison between different interpolations

Utilizing the measured distances, the eCley can measure the distances of objects and implement 3D position. Figure 6.8 shows the results of the application that measures the distances of objects exactly. The new subpixel distance measurement of stereo pixel pairs enables the eCley to measure the distances of objects effectively.

In the same scene, I used different approaches to attain the subpixel disparities. This gave rise to different results. As Figure 6.9 shows, the cubic interpolation outperforms the mean and bilinear interpolations. Clearly, a common tendency for all approaches is that the accuracy of measurement decreases with increasing distances of objects.

Chapter 7

Conclusions and Further Work

7.1 Conclusions

In many high-technology fields, such as informatics, artificial intelligence, aerospace, national defence, and so on, applied optical imaging systems are being used more and more widely. In the meantime, there is an increasing expectation of a miniaturization of the whole system. However, due to the limitations of the fabrication technology and the effects of optical diffraction, a minimization through modifying the size of the electronic elements and mechanical devices is faced with a dilemma. An alternative is to minimize the entire optical part of the system so as to reach the goal of a small volume and light weight for the whole imaging system.

The artificial compound eye is a novel branch of imaging systems, which combines bionics, integrated optics, and micro optics. Researches on artificial compound eyes can overcome the limitations of conventional cameras, such as their enormous size, narrow FOV (Field of View), and insensitivity to objects. Thus, their important practical value has attracted a lot of attention to research on artificial compound eyes. Quite successful applications include, for example, the imaging needed for precision-guided missiles, a polarization navigator for micro aerial vehicles, vision navigation and object recognition for robots, fingerprint and credit card identification for information security, full FOV imaging for small distance detection, and so on.

The electronic cluster eye (eCley) is a new artificial superposition compound eye with super-resolution. Its many outstanding characteristics, such as small volume, light weight, large FOV, short focal length, and high sensitivity, increase some potential abilities such as stereo vision in the same imaging system. Nowadays, applications based on the eCley mainly concentrate on the processing of 2D image information. When users want to obtain 3D image information, extra processing devices or sensors such as Time of Flight (TOF) depth sensors and laser scanners must be used to capture depth information. To get information on the depth of objects we can certainly also use passive image devices such as the stereo camera. Nevertheless, the cost of the extra devices is quite significant and their bulky volume conflicts with the goal of miniaturization, thus limiting its application. Taking aim at these problems, I can obtain some cues from the structure of eCley, which contains 17×13 microlens channels with the same optical parameters and geometry size, and then the image channels are retained straight on each row and column. These properties enable the functionality of stereo vision in the eCley. Just in this context, the present dissertation analyzes the significant fact that the transitional areas of image brightness contain rich information, including depths or distances of objects, in either the usual camera or the eCley. The advantage of the eCley is to allow a multi baseline stereo arrangement. Furthermore, the varying relation between the positions of feature points and their intensities in transitional areas can be used to determine the subpixel distances of the corresponding pixel pair in adjacent channels. I mainly develop theoretically a mathematical model of stereo vision in the eCley based on its special structure, discuss the optical correction and geometric calibration that are essential to high-precision measurements, study the implementation methods of the subpixel baselines for each pixel pair based on intensity information and gradient information in transitional areas, and implement a real-time subpixel distance measurement method for objects through the use of features on the edge. I also verify my proposed methods by using both an artificial synthetic stereo channel image and a large number of real images captured in diverse scenes, and discuss the experimental results with respect to various influencing factors.

The whole dissertation demonstrates my work in the following five parts.

- Firstly, I introduce the methods of distance measurement based on the canonical stereo camera system, develop a mathematical model for depth information, and analyze the limitations of conventional stereo matching algorithms for eCley due to its short focal length and miniaturized structure.

- Secondly, the ill-posed problem of image processing, especially for distance measurement, is demonstrated, followed by analyzing the structure of eCley and the significant characteristics of the transitional area between objects and background, e.g. that the variation of intensity reflects the position of a feature point. With reference to this fact, the whole framework of stereo vision in a multilens camera, such as the eCley, is illustrated, and how 3D depth information results from the implementation of subpixel distance measurement for objects.

- Afterwards, I propose two approaches to brightness correction: maximal mean and voted probability. The offsets of each adjacent two channels are also derived from the difference between the center coordinates of two neighboring channels, which are coarsely calculated by circle fitting and then more finely by straight line fitting. In addition, after analyzing the optical radial and tangential distortions, I create a projective model including various distortion and transformation parameters, and then employ the method of least squares to solve for the optimal parameters that maintain the suitability for distance measurement of the calibrated channel images.

- Then, I discuss the detection of subpixel features in the transitional areas as a ill-posed problem. Based on the characteristics of intensity and gradient, I get the subpixel positions of pixel points by the max mean and edge reconstruction, and determine the subpixel baseline between the two corresponding pixels in adjacent channels. After traversing the whole channel (left or right channel), all the baselines of pixel pairs in the pair of channels can be calibrated accurately at the subpixel level. The experimental results, involving both synthetic channel images and numerous real images shot in real scenes, verify the proposed methods. The factors of precision are discussed, for instance, the subpixel position versus error, the variation of illumination versus error, the objects' distance versus the precision, and so on.

- Finally, I propose a method of real-time subpixel distance measurement for objects, which removes noise and some uncertainty by a gaussian filter, and finds the subpixel positions by interpolating the characteristic curve in the transitional areas to determine the depth information of an object. To ensure overall efficiency, the whole algorithm adopts a coarse-to-fine strategy. I also compare the different accuracies arising from different interpolating approaches in my experiments, and analyze the sensitivities of the offset, distortion, and distance of the object.

In general, these proposed algorithms give eCley the ability to detect the 3D positions of objects, i.e. they implement a real-time subpixel distance measurement for objects. The main point is that the measurement range has been extended from 8 cm with conventional methods to about 80 cm. This allows wide field supervision e.g., in automotive applications.

7.2   Further Work

Although stereo vision is a novel function in the eCley, which greatly extends its applicability, there is still room for improvements in the accuracy and efficiency of real-time 3D measurements in future research.

- Although the approaches proposed in this dissertation calibrate the distortion and correct the brightness, some specific improvements are necessary. For example, the distortion at the boundary of the channel should be dealt with before computing the distance, since it may cause the aberrance in the properties in the transitional area. Furthermore, removing the effect of uneven illuminance in the test environment is quite important for obtaining precise results, something which limits its applicability: a more robust correction and calibration method should be developed.

- This dissertation employs both the intensity and gradient features to derive the subpixel positions of corresponding feature points, but the number of pixels in a 1D transitional range is only around four, which can only provide limited information for the construction of the characteristic curve. This paucity of points can not guarantee the accuracy of the reconstructed curve, so that the positions of the feature points are not quite accurate. How to utilize more information from related pixels or more pixels is to be considered in future research.

- In the real-time processing algorithm, the dissertation uses a preset size of the transitional area for a quick decision about the transitional area using coarse edges and interpolation to improve the efficiency. However, the Canny edge detector is used to get the coarse edge, which is computationally expensive. A simple edge detector with suitable accuracy and high efficiency should be adopted as part of the overall coarse-to-fine strategy, but the influence of this change had to be investigated.

- The uncertainty from other factors such as illumination could be analyzed in detail in order to see how deviations in experiment cause differences in the results.

References

BIRCHFIELD, S. & TOMASI, C. (1998). A pixel dissimilarity measure that is insensitive to image sampling. *Pattern Analysis and Machine Intelligence, IEEE Transactions on*, **20**, 401 –406. 16

BOBICK, A.F. & INTILLE, S.S. (1999). Large occlusion stereo. *International Journal of Computer Vision*, **33**, 181–200. 16

BOUCHARA, F., BERTRAND, M., RAMDANI, S. & HAYDAR, M. (2007). Sub-pixel edge fitting using b-spline. In *Proceedings of the 3rd international conference on Computer vision/computer graphics collaboration techniques*, MIRAGE'07, 353–364, Springer-Verlag, Berlin, Heidelberg. 31

BOUGUET, J.Y. (2010). Camera calibration toolbox for matlab. 52, 54

BRADSKI, G. & KAEHLER, A. (2008). *Learning OpenCV: Computer Vision with the OpenCV Library*. ISBN-10: 0596516134, O'REILLY. 54

BREDER, R., VIEIRA ESTRELA, V. & DE ASSIS, J. (2009). Sub-pixel accuracy edge fitting by means of b-spline. In *Multimedia Signal Processing, 2009. MMSP '09. IEEE International Workshop on*, 1–5. 31

BROWN, M., BURSCHKA, D. & HAGER, G. (2003). Advances in computational stereo. *IEEE Transactions on Pattern Analysis and Machine Intelligence*, **25**, 993 – 1008. 12, 13

BRUECKNER, A., A.CKNER, DUPARR, J., LEITEL, R., DANNBERG, P., BRUER, A. & TNNERMANN, A. (2010). Thin wafer-level camera lenses inspired by insect compound eyes. *Opt. Express*, **18**, 24379–24394. 2, 21, 32, 33

CHEN, G. & HONG YANG, Y. (1995). Edge detection by regularized cubic b-spline fitting. *Systems, Man and Cybernetics, IEEE Transactions on*, **25**, 636–643. 58

CYGANEK, B. & SIEBERT, J.P. (2009). *An Introduction To 3D Computer Vision Techniques And Algorithms*. John Wiley & Sons, Ltd. 11, 18, 23, 59, 62, 90

DUPARR, J., DANNBERG, P., SCHREIBER, P., BRÄUER, A. & TÜNNERMANN, A. (2005a). Thin compound-eye camera. *Applied Optics*, **44**, 2949–2956. 2

DUPARR, J., SCHREIBER, P., MATTHES, A., PSHENAY-SEVERIN, E., BRÄUER, A., TÜNNERMANN, A., VÖLKEL, R., EISNER, M. & SCHARF, T. (2005b). Microoptical telescope compound eye. *Optics Express*, **13**, 889–903. 1

DUPARR, J., RADTKE, D., BRCKNER, A. & BRUER, A. (2007). Latest developments in micro-optical artificial compound eyes: a promising approach for next generation ultracompact machine vision. 65030I–65030I–12. 4

DUPARR, J.W. & WIPPERMANN, F.C. (2006). Micro-optical artificial compound eyes. *Bioinspiration & Biomimetics*, **1**, R1–R16. 1, 2

FABIJANSKA, A. & SANKOWSKI, D. (2010). Edge detection with sub-pixel accuracy in images of molten metals. In *Imaging Systems and Techniques (IST), 2010 IEEE International Conference on*, 186–191. 29

FABIJASKA, A. (2012). A survey of subpixel edge detection methods for images of heat-emitting metal specimens. *International Journal of Applied Mathematics and Computer Science*, **22**, 695–710. 28, 65

FOROOSH, H. & BALCI, M. (2004). Sub-pixel registration and estimation of local shifts directly in the fourier domain. In *Image Processing, 2004. ICIP '04. 2004 International Conference on*, vol. 3, 1915–1918 Vol. 3. 27

FOROOSH, H., ZERUBIA, J. & BERTHOD, M. (2002). Extension of phase correlation to subpixel registration. *Image Processing, IEEE Transactions on*, **11**, 188–200. 27

GHOSAL, S. & MEHROTRA, R. (1993). Orthogonal moment operators for subpixel edge detection. *Pattern Recognition*, **26**, 295 – 306. 31

HARTLEY, R. & ZISSERMAN, A. (2004). *Multiple View Geometry in Computer Vision*. Cambridge University Press, ISBN: 0521540518, 2nd edn. 36, 47

HIRSCHMÜLLER, H., INNOCENT, P.R. & GARIBALDI, J. (2002). Real-time correlation-based stereo vision with reduced border errors. *Int. J. Comput. Vision*, **47**, 229–246. 14, 34

HORISAKI, R., IRIE, S., OGURA, Y. & TANIDA, J. (2007). Three-dimensional information acquisition using a compound imaging system. *Optical Review*, **14**, 347–350. 4, 20

HORISAKI, R., NAKAO, Y., TOYODA, T., KAGAWA, K., MASAKI, Y. & TANIDA, J. (2008). A compound-eye imaging system with irregular lens-array arrangement. In *Proc. SPIE 7072*, vol. 7072, 70720G–70720G–9. 20

HORISAKI, R., NAKAO, Y., TOYODA, T., KAGAWA, K., MASAKI, Y. & TANIDA, J. (2009). A thin and compact compound-eye imaging system incorporated with an image restoration considering color shift, brightness variation, and defocus. *Optical Review*, **16**, 241–246. 20

HUMENBERGER, M., ZINNER, C., WEBER, M., KUBINGER, W. & VINCZE, M. (2010). A fast stereo matching algorithm suitable for embedded real-time systems. *Comput. Vis. Image Underst.*, **114**, 1180–1202. 12

KAGAWA, K., TANAKA, E., YAMADA, K., KAWAHITO, S. & TANIDA, J. (2012). Deep-focus compound-eye camera with polarization filters for 3d endoscopes. *Proc. SPIE 8227*, 822714–822714–8. 4

KLAUS, A., SORMANN, M. & KARNER, K. (2006). Segment-based stereo matching using belief propagation and a self-adapting dissimilarity measure. In *Pattern Recognition, 2006. ICPR 2006. 18th International Conference on*, vol. 3, 15 –18. 17

KLETTE, R., SCHLUENS, K. & KOSCHAN, A. (1998). *Computer Vision. Three-Dimensional Data from Images*. Springer-Verlag. 62

KOLMOGOROV, V. (2004). *Graph based algorithms for scene reconstruction from two or more views*. Ph.D. thesis, Ithaca, NY, USA, aAI3114475. 16

KOWALCZUK, J., PSOTA, E. & PEREZ, L. (2013). Real-time stereo matching on cuda using an iterative refinement method for adaptive support-weight correspondences. *Circuits and Systems for Video Technology, IEEE Transactions on*, **23**, 94 –104. 15

LI, L. & YI, A.Y. (2010). Development of a 3d artificial compound eye. *Opt. Express*, **18**, 18125–18137. 4

LIU, C., XIA, Z., NIYOKINDI, S., PEI, W., SONG, J. & WANG, L. (2004). Edge location to sub-pixel value in color microscopic images. In *Intelligent Mechatronics and Automation, 2004. Proceedings. 2004 International Conference on*, 548–551. 29

LU, J., LAFRUIT, G. & CATTHOOR, F. (Jan. 2008). Anisotropic local high-confidence voting for accurate stereo correspondence. In *Proc. Image Processing SPIE-IS&T*, vol. 6812, 1–12. 14, 15

LUCAS, B.D. & KANADE, T. (1981). An iterative image registration technique with an application to stereo vision. In *Proceedings of the 7th international joint conference on Artificial intelligence - Volume 2*, IJCAI'81, 674–679, Morgan Kaufmann Publishers Inc., San Francisco, CA, USA. 27

LYVERS, E.P., MITCHELL, O.R., AKEY, M.L. & REEVES, A.P. (1989). Subpixel measurements using a moment-based edge operator. *IEEE Trans. Pattern Anal. Mach. Intell.*, **11**, 1293–1309. 29, 30

MACVICAR-WHELAN, P. & BINFORD, T. (1991). Line finding with subpixel precision. In *Proceedings of the DARPA Image Understanding Workshop*, 26–31, Washington, DC, USA. 29

MÜHLMANN, K., MAIER, D., HESSER, J. & MÄNNER, R. (2002). Calculating dense disparity maps from color stereo images, an efficient implementation. *Int. J. Comput. Vision*, **47**, 79–88. 26

NAKAMURA, T., HORISAKI, R. & TANIDA, J. (2012). Computational superposition compound eye imaging for extended depth-of-field and field-of-view. *Optics Express*, **20**, 27482–27495. 1, 2

NEFIAN, A.V., HUSMANN, K., BROXTON, M., TO, V., LUNDY, M. & HANCHER, M.D. (2009). A bayesian formulation for sub-pixel refinement in stereo orbital imagery. In *Proceedings of the 16th IEEE international conference on Image processing*, ICIP'09, 2337–2340, IEEE Press, Piscataway, NJ, USA. 27

NEHAB, D., RUSINKIEWICZ, S. & DAVIS, J. (2005). Improved sub-pixel stereo correspondences through symmetric refinement. In *Proceedings of the Tenth IEEE International Conference on Computer Vision (ICCV'05) Volume 1 - Volume 01*, ICCV '05, 557–563, IEEE Computer Society, Washington, DC, USA. 26

OTSU, N. (1979). A threshold selection method from gray-level histograms. *IEEE Transactions on Systems, Man and Cybernetics*, **9**, 62–66. 45

PEREZ, M., PAGLIARI, C. & DENNIS, T. (2005). A zero-crossing edge detector with improved localization and robustness to image brightness and contrast manipulations. In *Image Processing, 2005. ICIP 2005. IEEE International Conference on*, vol. 2, II–482–5. 29

PRATT, W. (2001). *Digital Image Processing, 3rd edn,.* John Wiley & Sons, Ltd. 59

PSARAKIS, E.Z. & EVANGELIDIS, G.D. (2005). An enhanced correlation-based method for stereo correspondence with sub-pixel accuracy. In *Proceedings of the Tenth IEEE International Conference on Computer Vision (ICCV'05) Volume 1 - Volume 01*, ICCV '05, 907–912, IEEE Computer Society, Washington, DC, USA. 27

RAO, C., TOUTENBURG, H., SHALABH & HEUMANN, C. (2008). *Linear Models and Generalizations : Least Squares and Alternatives.* ISBN 978-3-540-74226-5, Springer Berlin Heidelberg. 47

SCHARSTEIN, D. & SZELISKI, R. (2002). A taxonomy and evaluation of dense two-frame stereo correspondence algorithms. *International Journal of Computer Vision*, **47**, 7 –42. 13, 15

SCHMID, C. & ZISSERMAN, A. (2000). The geometry and matching of lines and curves over multiple views. *International Journal of Computer Vision*, **40**, 199–233. 12

SHI, C., WANG, G., PEI, X., BEI, H. & LIN, X. (2012). High-accuracy stereo matching based on adaptive ground control points. 17

SHIMIZU, M. & OKUTOMI, M. (2005). Sub-pixel estimation error cancellation on area-based matching. *Int. J. Comput. Vision*, **63**, 207–224. 26

STOLLBERG, K., BRUECKNER, A., DUPARR, P., J.AND DANNBERG & BRAEUER, A. (2009). The gabor superlens as an alternative wafer-level camera approach inspired by superposition compound eyes of nocturnal insects. *Optical Engineering*, **17**, 15747–1575. 2

SZELISKI, R. & SCHARSTEIN, D. (2002). Symmetric sub-pixel stereo matching. In *Proceedings of the 7th European Conference on Computer Vision-Part II*, ECCV '02, 525–540, Springer-Verlag, London, UK, UK. 26

TABATABAI, A.J. & MITCHELL, O.R. (1984). Edge location to subpixel values in digital imagery. *Pattern Analysis and Machine Intelligence, IEEE Transactions on*, **PAMI-6**, 188–201. 29, 30

TAPPEN, M. & FREEMAN, W. (2003). Comparison of graph cuts with belief propagation for stereo, using identical mrf parameters. In *Computer Vision, 2003. Proceedings. Ninth IEEE International Conference on*, 900 –906 vol.2. 16

TIKHONOV, A.N. & ARSENIN, V.Y. (1977). *Solutions of ill-posed problems.* Halsted. 58

TOMBARI, F., MATTOCCIA, S., DI STEFANO, L. & ADDIMANDA, E. (2008). Classification and evaluation of cost aggregation methods for stereo correspondence. In *Computer Vision and Pattern Recognition, 2008. CVPR 2008. IEEE Conference on*, 1 –8. 15

TONG, G., LIU, R. & TAN, J. (2011). 3d information retrieval in mobile robot vision based on spherical compound eye. In *2011 IEEE International Conference on Robotics and Biomimetics (ROBIO)*, 1895 –1900. 1

TRUCCO, E. & VERRI, A. (1998). *Introductory Techniques for 3-D Computer Vision*. Prentice Hall. 54

VEKSLER, O. (2003). Fast variable window for stereo correspondence using integral images. In *Computer Vision and Pattern Recognition, 2003. Proceedings. 2003 IEEE Computer Society Conference on*, vol. 1, I: 556–561. 14

VENKATESWAR, V. & CHELLAPPA, R. (1995). Hierarchical stereo and motion correspondence using feature groupings. *International Journal of Computer Vision*, **15**, 245–269. 12

VOELKEL, R., HERZIG, H.P., NUSSBAUM, P., DANDLIKER, R. & HUGLE, W.B. (1996). Microlens array imaging system for photolithography. *Optical Engineering*, **35**, 3323–3330. 1

XIANWEI, G., WEI-XING, Y., HONG-XING, Z., ZHEN-WU, L., QING, S. & HONG-HAI, S. (2013). Progress in design and fabrication of artificial compound eye optical systems. *Chinese Optics*, **6**, 34–45. 4

XU, G.S. (2009). Sub-pixel edge detection based on curve fitting. In *Proceedings of the 2009 Second International Conference on Information and Computing Science - Volume 02*, ICIC '09, 373–375, IEEE Computer Society, Washington, DC, USA. 29

YANG, Q., WANG, L., YANG, R., STEWENIUS, H. & NISTER, D. (2009). Stereo matching with color-weighted correlation, hierarchical belief propagation, and occlusion handling. *Pattern Analysis and Machine Intelligence, IEEE Transactions on*, **31**, 492 –504. 17, 26

YAO, Y. & JU, H. (2009). A sub - pixel edge detection method based on canny operator. In *Proceedings of the 6th international conference on Fuzzy systems and knowledge discovery - Volume 5*, FSKD'09, 97–100, IEEE Press, Piscataway, NJ, USA. 31, 83

YOON, K.J. & KWEON, I.S. (2005). Locally adaptive support-weight approach for visual correspondence search. In *Computer Vision and Pattern Recognition, 2005. CVPR 2005. IEEE Computer Society Conference on*, vol. 2, 924 – 931. 15

YU, W. & XU, B. (2009). A sub-pixel stereo matching algorithm and its applications in fabric imaging. *Mach. Vision Appl.*, **20**, 261–270. 27

YUANYUAN, T., QINGCHANG, T., HAIBO, Z., ZHIBIAO, S. & SHENGQUAN, H. (2010). Stability analysis of subpixel edge location algorithm. *Chinese Journal of Computer Engineering*, **36**, 211–213. 31

YUEYING, Z., WEIDE, Q., JUN, L. & SHIFAN, Z. (2007). Study on sub-pixel edge location method based on curve fitting. *Optical Technique*, **32**, 386–389. 31

YUILLE, A.L. (1986). Scaling theorems for zero crossings. *IEEE Transactions on Pattern Analysis and Machine Intelligence*, **8**, 15–25. 57

ZHANG, K., LU, J. & LAFRUIT, G. (2009a). Cross-based local stereo matching using orthogonal integral images. *Circuits and Systems for Video Technology, IEEE Transactions on*, **19**, 1073 –1079. 14

ZHANG, K., LU, J., YANG, Q., LAFRUIT, G., LAUWEREINS, R. & VAN GOOL, L. (2011). Real-time and accurate stereo: A scalable approach with bitwise fast voting on cuda. *Circuits and Systems for Video Technology, IEEE Transactions on*, **21**, 867 –878. 14, 15

ZHANG, W.J., LI, D. & YE, F. (2009b). Sub-pixel edge detection method ased on sigmoid function fitting. *Journal of South China University of Technology*, **37**, 39–43. 65

ZHANG, Z. (2000). A flexible new technique for camera calibration. *IEEE Transactions on Pattern Analysis and Machine Intelligence*, **22**, 1330–1334. 50, 51, 52

ZHONGHAI, H., BAOGUANG, W. & YIBAI, L. (2002). Study of method for generating ideal edges. *Optics and Precision Engineering*, **10**, 89–93. 67